Python 自动化测试实战

鹿瑞峰 编著

北京大学出版社

内 容 简 介

本书的写作初衷是为了帮助更多功能测试人员转型自动化测试方向。在转型过程中，主流自动化测试技术和应用场景的结合是非常重要的一环。本书从自动化测试理论入手，全面地阐述自动化测试的意义及实施过程。全文以Python语言驱动，结合真实案例分别对主流自动化测试工具Selenium、Robot Framework、Postman、Python+Requests、Appium等进行系统讲解。通过学习本书，读者可以快速掌握主流自动化测试技术，并帮助读者丰富测试思维，提高Python编码能力。

本书实用性强，不仅是转型自动化测试方向的一本案头书，也是一本特别好用、实用的操练手册。

图书在版编目(CIP)数据

Python自动化测试实战 / 鹿瑞峰编著. —北京：北京大学出版社，2019.12
ISBN 978-7-301-30910-0

Ⅰ.①P… Ⅱ.①鹿… Ⅲ.①软件工具－程序设计 Ⅳ.①TP311.561

中国版本图书馆CIP数据核字（2019）第238682号

书　　　名	**Python自动化测试实战** PYTHON ZIDONGHUA CESHI SHIZHAN
著作责任者	鹿瑞峰　编著
责任编辑	吴晓月　王继伟
标准书号	ISBN 978-7-301-30910-0
出版发行	北京大学出版社
地　　　址	北京市海淀区成府路205 号　100871
网　　　址	http://www.pup.cn　新浪微博:@北京大学出版社
电子信箱	pup7@pup.cn
电　　　话	邮购部 010-62752015　发行部 010-62750672　编辑部 010-62570390
印刷者	北京溢漾印刷有限公司
经销者	新华书店
	787毫米×1092毫米　16开本　18.5印张　420千字 2019年12月第1版　2021年4月第3次印刷
印　　　数	6001-9000册
定　　　价	69.00元

未经许可，不得以任何方式复制或抄袭本书之部分或全部内容。
版权所有，侵权必究
举报电话: 010-62752024　电子信箱: fd@pup.pku.edu.cn
图书如有印装质量问题，请与出版部联系，电话: 010-62756370

前言

市场上关于自动化测试方面的书有很多,如专门介绍 Selenium、Robot Framework 及其他常用测试工具类(如 Git、Jenkins、Postman 等)的书籍。但唯独缺少一本可以将主流自动化测试技术与测试工具整合在一起的书。于是,《Python 自动化测试实战》应运而生。

本书深入浅出,并结合众多工作中的实例引领读者从功能自动化测试到接口自动化测试方面的实战。本书中的案例是结合作者多年的知识积累和实务工作经验而设计的。在对自动化测试技术阐述的同时,图文并茂,没有太多理论条文,避免实操过程中的枯燥、乏味。

书的写作类型一般可以分为两种,一种是有思想的书,另一种是工具类的书。在笔者看来,本书更像是一本整合目前主流自动化测试工具的教程书。本书会告诉读者如何使用自动化测试工具,并将自动化测试技术应用到实际工作中,最终解决一些测试过程中遇到的问题。

本书从自动化测试理论入手,以 Python 语言为驱动核心,围绕 Selenium、Robot Framework 等自动化测试工具在功能自动化测试领域中的实际应用而展开。此外,本书还系统地介绍了接口自动化测试在企业中实施的必要性,使用 Requests 库、Postman、Git、Charles、Jenkins 等测试工具并结合大量案例来演示接口测试在实际过程中的应用。

受作者水平和成书时间所限,本书难免存有疏漏和不当之处,敬请读者指正。此外,感谢出版社编辑王继伟一路以来的耐心指导。

本书特色

1. 特别注重实操性,可快速上手

本书内容涵盖了 Selenium、Robot Framework、Postman、Charles、Python+Requests、Git、Jenkins 等自动化测试工具所必须掌握的知识,从内容结构上非常注重知识的实用性和可操作性。本书在对知识点进行介绍时,拒绝深奥的理论条文,强调实务操作,使读者快速上手。本书由浅入深、循序渐进的知识讲授,完全遵循和尊重初学者对自动化测试技术知识的认知规律。

2. 图文并茂，大量案例引导全程

本书图文并茂，通过大量实际工作中的案例来引导读者学习并实操。不管读者是否为零基础，都可以按照教程在短时间内上手。本书更像是一本自动化测试技能字典，所有案例源码都可以在 GitHub 上下载，还可以通过查阅的方式来获取答案来验证自己的学习成果。

3. 涉及自动化测试工具范围广，可通过本书学习多种测试技能

本书是目前市场上涉及自动化测试工具较为广泛的一本书，涵盖了主流功能自动化测试工具（如 Selenium、Robot Framework）、主流接口自动化测试工具（如 Postman、Python+Requests），以及测试辅助工具（如 Charles、Git、Jenkins 等）。笔者相信本书是功能测试人员转型为自动化测试技术方向的一本必不可少的书。

本书内容及知识体系

第 1 章 自动化测试理论

本章引领读者了解实施自动化测试的意义和自动化测试实施的过程，为后续学习自动化测试技术抛砖引玉。

第 2 章 Selenium 3 自动化测试实战

本章涵盖目前主流功能自动化测试工具 Seleneium 3 的应用，讲解了 Selenium 3 自动化测试环境搭建，并结合大量案例来演示多种 Webdriver API 在自动化测试中的应用、unittest 单元测试框架、数据驱动测试（DDT）实战及自动化脚本的数据分离和封装设计实战等，在最后通过一个实战项目讲解如何从 0 到 1 搭建一个丰富且完整的功能自动化测试框架（包含用例组织、数据封装、配置文件管理、日志跟踪、自动生成邮件和持续集成等）。

第 3 章 Robot Framework 自动化测试实战

本章以 Python 语言为驱动，系统讲解了 Seleneium2Library 库及 DatabaseLibrary 库在自动化测试中的应用实战，其中包含 Selenium2Library 库和 DatabaseLibrary 库中常用关键字使用案例解析、关键字封装、项目分层设计与开发实战等，最后结合 Jenkins 实现持续集成，输出自动化测试报告。掌握本章后，读者可以独立开展自动化测试任务。

第 4 章 接口测试基础

本章以接口测试理论为切入点，系统地讲解了接口的定义、接口测试流程及接口测试用例设计原则和注意事项，重点讲解了 HTTP 及在接口测试中的重要性。在本章最后阐述了接口测试工作中绕不开的话题，即 Cookie 和 Session 的工作原理。

第 5 章 Charles 抓包工具实战

本章系统地讲解了主流抓包工具 Charles 在实际工作中的应用。通过本章，读者可以掌握 Charles 工具的配置、设置代理、计算机端抓包设置、手机端抓包设置、Charles 常见问题解析，以

及 Charles 在接口测试中的应用等，了解 Charles 抓包工具在接口自动化测试中的重要性。

第 6 章 Postman 接口测试高级实战

本章以 Postman 接口测试工具（Collections、Runner、Code、Globals 等）核心功能为出发点，重点讲解如何使用 Postman 处理 HTTP 请求和 WebServices 请求案例，以及在接口测试过程中如何设置接口断言、处理数据转换（JSON 和 XML）、解决动态参数的依赖和调用、分离公共数据、转换多种编程语言测试脚本等。本章最后通过 Newman+Jenkins+Postman 实现持续集成，可以定时接口自动化测试任务。

第 7 章 Python 接口自动化测试实战

本章系统地讲解了使用 Requests 模块实现接口自动化测试。从 Requests 模块常用示例（GET、POST、JSON、Requests Headers、Response 等）入手，围绕 Cookie、Session、Token、上传文件、序列化和反序列化等案例开展接口测试。最后结合 Python 3+Requests+unittest 讲解接口自动化测试框架的设计和开发思路。

第 8 章 Robot Framework 接口自动化实战

本章讲解了 Collections 库、ExcelLibrary 库、RequetsLibrary 库中常见关键字在接口测试中的使用。此外，还涉及如何使用 ExcelLibrary 库进行测试数据的维护和管理、封装并调用接口关键字，以及测试数据与业务分离等。掌握本章后，读者可以使用 Robot Framework 工具开展接口自动化测试任务。

第 9 章 Appium 自动化测试实战

本章系统地讲解了 Appium 移动端自动化测试环境的搭建，结合丰富的案例，基于 Android 操作平台演示 WebDriver 在实际工作中的应用，主要内容包括自动化截图实战、滑动实战、多点触控实战、键盘事件、等待函数实战等及使用 Appium 测试框架完成一个自动化项目实战脚本的演示（涵盖需求分析→自动化用例设计→脚本编写→脚本封装重构→测试案例运行及分析等）。掌握本章后，读者可以独立开展移动端自动化测试任务。

第 10 章 Git 版本控制工具实战

本章通过实际案例来演示 Git 在项目管理中的应用，包括版本库的创建、文件的提交和跟踪管理，以及版本回退等。此外，还涉及 GitHub 的配置及远程库的添加和远程克隆等操作。随着企业内部技术的不断升级和调整，相信掌握 Git 工具的使用方法是测试人员的必备技能之一。

本书读者对象

- 刚入行软件测试的测试人员
- 想转行从事接口自动化测试的测试人员
- 想转行从事功能自动化的测试人员
- 想学习测试框架开发的测试人员

- 想转行从事 Python 自动化方向的测试人员
- 想学习多种自动化测试工具的测试人员

资源下载

本书所涉及的源代码及软件的下载地址已上传到百度网盘,供读者下载。请读者关注封底"博雅读书社"微信公众号,找到"资源下载"栏目,根据提示获取。

第1章 自动化测试理论

1.1 自动化测试现状 ··· 2

1.2 自动化测试的定义 ··· 2

1.3 自动化测试流程 ··· 3

1.4 自动化测试用例编写 ··· 4

第2章 Selenium 3 自动化测试实战

2.1 搭建自动化环境 ··· 6

    2.1.1 搭建 Python 3 环境 ·· 6

    2.1.2 搭建 Selenium 3 自动化环境 ······································ 8

2.2 配置浏览器驱动 ··· 9

    2.2.1 配置 IE 浏览器驱动 ·· 9

    2.2.2 配置 Firefox 浏览器驱动 ··· 10

    2.2.3 配置 Chrome 浏览器驱动 ··· 11

2.3 元素定位实战 ··· 12

    2.3.1 单个元素定位实战 ··· 12

2.3.2 多个元素定位实战 15
2.3.3 使用 By 类定位 17
2.3.4 使用 JavaScript 定位 17
2.3.5 使用 JQuery 定位 19

2.4 下拉框实战 19
2.4.1 value 属性定位 20
2.4.2 index 属性定位 20
2.4.3 visible_text 属性定位 21
2.4.4 元素二次定位实战 21

2.5 鼠标操作实战 22
2.5.1 鼠标指针悬停实战 22
2.5.2 鼠标右键实战 23
2.5.3 鼠标双击实战 24

2.6 元素等待实战 24
2.6.1 强制等待 24
2.6.2 隐式等待 24
2.6.3 显示等待 25

2.7 表单切换实战 25
2.7.1 单表单切换实战 26
2.7.2 嵌套表单切换实战 26
2.7.3 平行表单切换实战 27
2.7.4 表单特殊情况处理 28

2.8 窗口切换实战 28
2.8.1 GET 方法实战 28
2.8.2 SWITCH 方法实战 29

2.9 警告框实战 30

2.10 JavaScript 实战 32
2.10.1 处理富文本实战 32
2.10.2 处理隐藏元素实战 33
2.10.3 处理 readonly 属性实战 33
2.10.4 处理浏览器滚动条实战 34

2.11 unittest 单元测试框架 ······ 36
- 2.11.1 unittest 简介 ······ 36
- 2.11.2 前置和后置 ······ 37
- 2.11.3 常用断言方法 ······ 38
- 2.11.4 setUpClass() 和 tearDownClass() ······ 39
- 2.11.5 测试固件分离实战 ······ 40
- 2.11.6 生成 HTML 测试报告 ······ 41

2.12 数据驱动测试实战 ······ 42
- 2.12.1 DDT 简介与安装 ······ 43
- 2.12.2 DDT 在自动化测试中的应用 ······ 43
- 2.12.3 Excel 自动化测试实战 ······ 46
- 2.12.4 Excel 整合 DDT 自动化测试实战 ······ 48
- 2.12.5 YAML 自动化测试实战 ······ 50
- 2.12.6 parameterized 参数化实战 ······ 52

2.13 发送邮件实战 ······ 53
- 2.13.1 纯文本的邮件实战 ······ 54
- 2.13.2 带附件的邮件实战 ······ 55
- 2.13.3 yagmail 发送邮件实战 ······ 57

2.14 自动化测试封装实战 ······ 58
- 2.14.1 自动化封装实战（一）······ 58
- 2.14.2 自动化封装实战（二）······ 60

2.15 测试框架封装和脚本的分层设计 ······ 62
- 2.15.1 PageObject 设计模式 ······ 62
- 2.15.2 分离测试固件 ······ 66
- 2.15.3 分离测试数据 ······ 68
- 2.15.4 用例失败截图 ······ 70
- 2.15.5 日志跟踪收集 ······ 72
- 2.15.6 生成 HTML 格式的测试报告 ······ 76
- 2.15.7 发送带附件的测试报告 ······ 77
- 2.15.8 项目持续集成 ······ 79
- 2.15.9 Jenkins+Allure 配置自动化测试报告 ······ 85

2.16 自动化测试扩展应用实战 ······ 90

2.16.1　配置 Firefox 无头模式 91
2.16.2　配置 Chrome 无头模式 91
2.16.3　多线程调用浏览器运行实战 91
2.16.4　搭建 PyCharm IDE 开发环境 93

第 3 章　Robot Framework 自动化测试实战

3.1　搭建 Robot Framework 环境 98

3.1.1　安装 Python 98
3.1.2　安装 wxPython 98
3.1.3　安装 Robot Framework 99
3.1.4　安装 RIDE 99
3.1.5　验证测试环境 100
3.1.6　制作 RIDE 快捷图标 101

3.2　安装与导入 Selenium2Library 库 101

3.2.1　安装 Selenium2Library 库 102
3.2.2　导入 Selenium2Library 库 102

3.3　浏览器驱动配置 103

3.3.1　配置 Firefox 浏览器驱动 103
3.3.2　配置 Chrome 浏览器驱动 104

3.4　元素定位实战 105

3.4.1　引用 id 105
3.4.2　引用 name 106
3.4.3　引用 link 107
3.4.4　引用 css 108
3.4.5　引用 xpath 109
3.4.6　xpath 定位动态属性 110

3.5　JQuery 定位实战 110

3.5.1　处理特殊单击事件 111
3.5.2　移除 readOnly 属性 112
3.5.3　处理 Display 隐藏元素 112

3.5.4　使用 JQuery 处理蒙层事件 ·· 114
  3.5.5　使用 JQuery 获取文本框中的值 ·· 114
  3.5.6　使用 JQuery 向文本框中输入内容 ····································· 114
3.6　获取窗口标题 ··· 115
3.7　获取文本信息 ··· 116
3.8　鼠标指针悬停实战 ··· 116
3.9　操作多窗口实战 ·· 117
3.10　操作下拉列表框实战 ··· 119
3.11　操作警告框实战 ·· 120
3.12　获取结果断言 ·· 120
3.13　项目执行顺序 ·· 123
3.14　常见问题整理 ·· 124
3.15　自定义关键字 ·· 125
3.16　参数化关键字 ·· 127
3.17　关键字驱动测试 ·· 129
3.18　Settings 界面简介 ·· 130
3.19　项目分层设计与开发实战 ·· 130
  3.19.1　构建操作动作关键字 ·· 131
  3.19.2　构建操作界面关键字 ·· 131
  3.19.3　构建操作流程关键字 ·· 132
  3.19.4　构建自动化测试用例 ·· 132
3.20　连接 MySQL 数据库实战 ··· 133
  3.20.1　安装与导入 DatabaseLibrary 库 ·· 133
  3.20.2　DatabaseLibrary 库中常用关键字 ·· 134
  3.20.3　连接 MySQL 数据库设置（一）·· 135
  3.20.4　连接 MySQL 数据库设置（二）·· 137
  3.20.5　基于 MySQL 数据库的实战 ··· 137
3.21　Jenkins+Robot Framework 持续集成 ··· 139
  3.21.1　安装 robot.hpi 插件 ··· 139

3.21.2 构建自由风格的任务 ·· 139
3.21.3 构建后操作界面设置 ·· 140
3.21.4 查看 Robot Framework 测试报告 ·· 140

第 4 章 接口测试基础

4.1 接口测试的定义 ··· 142
4.2 接口测试的目的 ··· 142
4.3 接口测试原理 ··· 143
4.4 接口测试流程 ··· 143
4.5 接口测试用例设计 ··· 144
4.6 HTTP 基础 ··· 145
  4.6.1 HTTP 请求报文 ·· 145
  4.6.2 HTTP 响应报文 ·· 146
  4.6.3 HTTP 状态码 ·· 147
  4.6.4 HTTP 请求方法 ·· 148
4.7 Cookie 和 Session ··· 148
  4.7.1 Cookie 的工作原理 ·· 148
  4.7.2 Session 的工作原理 ·· 149

第 5 章 Charles 抓包工具实战

5.1 下载与安装 Charles ·· 151
5.2 计算机端抓包设置 ··· 151
5.3 手机端抓包设置 ··· 155
5.4 Charles 过滤请求 ··· 157
5.5 Charles 常见问题 ··· 158

第 6 章 Postman 接口测试高级实战

6.1 安装 Postman ··· 160
6.2 Collections 简介 ·· 160
6.3 基于 HTTP 接口实战 ·· 161
    6.3.1 处理原生 Form ·· 162
    6.3.2 处理 JSON ·· 163
    6.3.3 处理 Cookie ·· 164
    6.3.4 处理 Session ··· 165
6.4 基于 Web Services 接口实战 ·· 168
    6.4.1 Web Services 接口示例 ·· 168
    6.4.2 引用 JavaScript 断言策略 ·· 170
    6.4.3 解决动态参数获取 ·· 171
    6.4.4 解决上下游动态参数依赖 ··· 172
    6.4.5 Collections 公共数据分离 ·· 174
    6.4.6 批量运行 Collections ·· 175
    6.4.7 使用 Newman+Jenkins 构建接口自动化任务 ················· 177

第 7 章 Python 接口自动化测试实战

7.1 安装 Requests 库 ··· 182
7.2 Requests 发送请求与参数传递 ··· 183
    7.2.1 定义请求样式 ·· 183
    7.2.2 发送 GET 请求 ·· 183
    7.2.3 发送 Form 表单 ·· 184
    7.2.4 发送 XML 数据 ··· 185
    7.2.5 发送 JSON 请求 ·· 186
7.3 处理 Token ··· 187
7.4 处理 Cookie ·· 189
7.5 处理 Session ··· 190

7.6	处理超时等待	193
7.7	Response 对象解析	194
7.8	Requests 文件上传实战	195
7.9	Requests 常见异常	197
7.10	序列化和反序列化	197
7.11	XML 与 JSON 数据之间的转换	199
7.12	接口测试框架设计和开发	201

- 7.12.1 测试框架简介 ... 201
- 7.12.2 重构 Requests 请求 ... 202
- 7.12.3 重构接口案例 ... 203
- 7.12.4 动态参数写入文件并读取 ... 204
- 7.12.5 处理接口上下游参数依赖 ... 205
- 7.12.6 重构 Excel 工具类 ... 206
- 7.12.7 动态参数赋值调用 ... 208
- 7.12.8 日志管理功能 ... 209
- 7.12.9 配置文件功能 ... 212
- 7.12.10 发送接口测试报告 ... 213

第 8 章 Robot Framework 接口自动化实战

8.1	Collections 库案例实战	217
8.2	ExcelLibrary 库案例应用	221
8.3	RequestsLibrary 库案例实战	225
8.4	ExcelLibrary 库数据管理案例实战	228

第 9 章 Appium 自动化测试实战

| 9.1 | 安装 Appium 环境 | 234 |

- 9.1.1 安装 Node.js 环境 ... 234

9.1.2 安装 Appium 工具 ········ 235
9.1.3 安装 Java 环境 ········ 237
9.1.4 安装 Android SDK ········ 239
9.1.5 安装 Android SDK Platform-Tools ········ 240
9.1.6 安装 Appium 工具 Client ········ 241
9.1.7 Appium 连接模拟器和真机 ········ 241
9.1.8 获取 APP 包名和 AppActivity ········ 242
9.1.9 Appium 第一个自动化脚本 ········ 243

9.2 使用 Monitor 定位元素 ········ 244
9.2.1 id 定位 ········ 245
9.2.2 name 定位 ········ 246
9.2.3 class 定位 ········ 247
9.2.4 xpath 定位 ········ 248
9.2.5 accessibilty_id 定位 ········ 249
9.2.6 android_uiautomator 定位 ········ 249

9.3 Native App 实战 ········ 250
9.3.1 模拟键盘事件 ········ 250
9.3.2 滑动封装实战 ········ 253
9.3.3 多点触控实战 ········ 255
9.3.4 自动化异常截图 ········ 256

9.4 Appium 完整脚本实战 ········ 257
9.4.1 测试需求分析 ········ 257
9.4.2 测试用例设计 ········ 258
9.4.3 测试脚本编写 ········ 259
9.4.4 测试结果及分析 ········ 260

9.5 Appium 常见问题 ········ 262

第 10 章 Git 版本控制工具实战

10.1 搭建 Git 环境 ········ 265
10.2 Git 基本操作 ········ 267

- 10.2.1 创建版本库 ………………………………………………………… 268
- 10.2.2 添加文件 …………………………………………………………… 268
- 10.2.3 文件跟踪管理 ……………………………………………………… 270
- 10.2.4 版本回退 …………………………………………………………… 272

10.3 Git 项目管理 …………………………………………………………… 274

- 10.3.1 配置 GitHub ……………………………………………………… 274
- 10.3.2 添加远程库 ………………………………………………………… 276
- 10.3.3 克隆远程库 ………………………………………………………… 279

第1章 自动化测试理论

本章从自动化测试现状、自动化测试的定义、自动化测试流程及自动化测试用例编写4个方面来阐述在开展自动化测试任务前，需要我们了解的自动化测试理论知识。通过本章，可以方便读者在后续的章节学习过程中将本章自动化测试理论知识与自动化测试工具有效结合，从而应用在日常工作中。

1.1 自动化测试现状

随着互联网行业的不断发展，互联网产品同质化现象越来越严重，用户对产品的使用更加追求细腻，用户体验也变得越来越重要。这就不得不强调在实际开发过程中软件质量的重要性。近年来，自动化测试技术一直不断被热捧，好像随便去一家公司面试都会被问到会不会自动化测试，到了企业是否真正运用另当别论。不难发现，自动化测试技术已经成为软件测试从业者必须掌握的一项技能。

可能很多人会认为测试思维比较重要，而测试工具或测试框架只是辅助提升效率。随着互联网产品的竞争日益激烈，企业要做更好的产品来服务用户，必须以技术来驱动创新。就像自动化测试技术的出现一样，其帮助人类从手工测试的重复且机械性的工作中解脱出来，从而把更多精力放在更有意义的事情上面。

这个社会唯一不会改变的事情就是一直在变。对于测试人员来说，技术需要不断更新，在实际工作中，要不断地提升自身技能，学习新的测试技术，这样才可以保证自身价值，不会被社会淘汰。

对于大多数进入软件测试领域的测试工程师而言，每日的测试任务基本都是重复且机械性的。长此以往，测试人员自身技能会怎么样？未来的职业发展又会是怎样？

软件测试技术是一个比较综合且复杂的技术学科，它涉及的知识面和涵盖的领域众多。绝大多数的测试工程师有很长一段时间几乎都停留在功能测试阶段。在这种情况下，对于一些刚入行不久的测试工程师来说，从功能测试转型到自动化测试或许是一个不错的选择。

同样地，自动化测试技术也是互联网行业发展的一种必然趋势，如果你想走得更远，可以向测试开发方向去努力。笔者相信，只要努力，就可以走得更远。

1.2 自动化测试的定义

自动化测试可以理解为通过测试脚本来替代手工测试（功能测试）的过程，它可以自动执行那些在正式的测试过程中已经存在的、重复但必需的任务，或者执行一些通过手工难以执行的附加测试。自动化测试对于持续交付和持续测试非常关键。

自动化测试实际上涵盖多种类型，如功能（黑盒）自动化测试（常见工具包括 Selenium、Robot Framework 等）、功能（白盒）自动化测试（如 Python 语言中的 Pyunit、Pytest 单元测试框架或 Java 语言中的 TestNG、Junit 单元测试框架等）。此外，还包括性能测试（LoadRunner、Jmeter、AB）和安全测试等。

1.3 自动化测试流程

在开展自动化测试任务前，必须要了解并掌握实施自动化测试的流程。如果连自动化测试流程都无法掌握，那么也就无法顺利开展自动化测试任务。自动化测试流程和功能测试的测试流程大同小异。本节内容结合笔者自身多年自动化测试工作经验总结得出，可能并不一定就是标准的自动化测试流程，但具备一定的参考价值。

1. 需求分析

拿到一个项目以后，首先要分析系统哪些模块适合做自动化测试，而且要了解清楚这部分自动化测试实施的意义是什么，能够为我们带来什么样的好处和价值。如果上来就盲目地开展自动化测试，那么结果一定是没有意义的。结合笔者自身多年测试工作经验，一般来说，在评估一个项目是否满足做自动化测试的条件时，可以从以下但不局限于这几个方面考虑：需求不会频繁变更、界面稳定、手工测试稳定、回归测试验证使用较为频繁的功能。

2. 测试方案选择

测试方案选择可以理解为测试工具的选择。在开展自动化测试任务时，一般会考虑选择什么样的方式来实现。最常用的就是采用工具录制或编码的方式来进行，笔者更推荐后者。通过编码可以实现更多开发者自定义的功能，尤其是当测试案例比较多时，要考虑维护、测试数据管理等问题。目前比较主流的自动化测试框架有 Selenium、Robot Framework、Appium、Roboium 和 MonkeyRunner 等，在本书中分别会有 Selenium、Robot Frameowork 和 Appium 等测试框架的案例介绍。此外，测试方案的选择还要结合项目团队所使用的开发语言、项目架构和项目环境等方面考虑。

3. 测试环境准备

开展自动化测试前，测试环境的准备是必需的。测试环境一般包括工具安装环境和自动化测试环境，如果使用 Selenium、Appium 开展自动化测试，就必须要安装语言环境，如安装 Python、Java，还要配置 IDE、JDK 等。不仅如此，还要考虑持续集成环境和版本管理等。

4. 测试框架设计

当编写了大量的自动化测试案例后，测试案例的维护会变得困难。对于测试案例中使用到的公共部分，如测试数据、配置文件和日志文件等都需要存放在不同的文件夹以便管理。因为在实际开发中，测试项目往往是由一个团队来完成的，不同的人会负责不同的测试模块，在后期还会涉及模块和模块间的相互调用等。

5. 测试用例执行

在自动化测试案例执行结束后，可以通过测试报告得知：本次测试执行了多少用例、通过数是多少、失败数是多少及失败原因是什么。还可以通过 Jenkins 定时构建自动化任务，如果涉及代码提交或更新，就会触发自动构建自动化任务，生成测试结果，最后将测试结果发送给相关负责人。

1.4 自动化测试用例编写

自动化测试用例和功能测试用例的区别不是很大，包含用例 ID、模块、前提条件、操作步骤、预期结果和实际结果等。下面来看一个登录功能的自动化测试用例，如图 1.1 所示。

用例ID	模块	前提条件	操作步骤	预期结果	实际结果
Login_001	登录	环境正常，系统稳定 存在已注册的账号： admin 存在已注册的密码： 123456	1.打开测试网址：http://mail.qq.com 2.在元素为ID=username的文本框内输入用户名：admin 2.在ID为password的文本框内输入密码：123456 3.单击ID为submit的登录按钮	1.跳转到登录界面 2.用户名和密码输入正确 3.登录成功，页面跳转	

图 1.1　登录功能的自动化测试用例

通过图 1.1，对比手工（功能）测试用例不难发现，自动化测试用例对测试步骤要求更加细致、准确，对测试数据的要求也更加完备、可靠。因为自动化测试本质是模拟用户的手工测试过程。

当然，自动化测试用例的编写还有其他方式，上述用例只是笔者在过往的工作中的一些经验分享。可能很多公司设计自动化测试用例和转换自动化测试用例的并非同一个人，所以要保证编写的自动化测试用例一定要方便其他人来编码自动化测试脚本。

第2章 Selenium 3自动化测试实战

目前 Web 端流行度最高、使用人数最多的自动化测试工具无疑是 Selenium。Selenium 工具优势众多，如支持多种编程语言、多平台、多种浏览器和并行执行测试用例等。掌握并使用 Selenium 工具是每一个测试工程师必须具备的技能。

2.1 搭建自动化环境

在开始 Selenium 工具实战前，首先要安装好 Python 环境。本节的 Python 环境是基于最新的 Python 3 来演示的，选择的版本是 Python 3.6.4。

2.1.1 搭建 Python 3 环境

访问 Python 官网（https://www.python.org），选择对应的平台进行下载。在安装 Python 时，要注意区分操作系统的版本是 32 位还是 64 位，如图 2.1 所示。

图 2.1　Python 安装界面

安装时注意选中"Add Python 3.6 to PATH"复选框，然后选择"Customize installation"自定义安装，会出现如图 2.2 所示的设置界面。在此界面不需要设置任何选项，直接单击"Next"按钮即可。

如图 2.3 所示，需要指定安装 Python 的路径，将 Python 的路径设置在 C:\ 根目录下。Python 安装完成后会自动创建一个 Python36 的目录。

单击"Install"按钮开始安装 Python，图 2.4 显示的是安装加载界面。

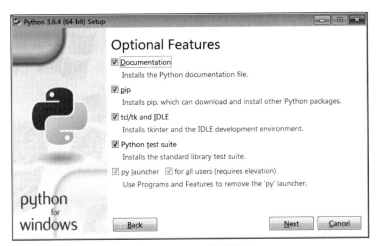

图 2.2 Python 设置界面

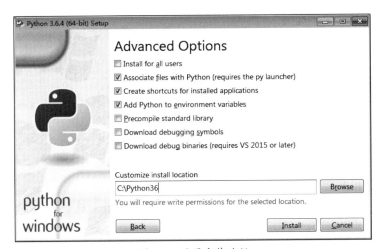

图 2.3 设置安装路径

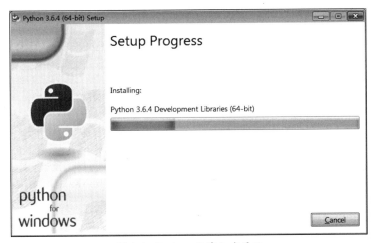

图 2.4 Python 安装加载界面

安装完成后，打开 C 盘根目录，查看 Python 目录，如图 2.5 所示。

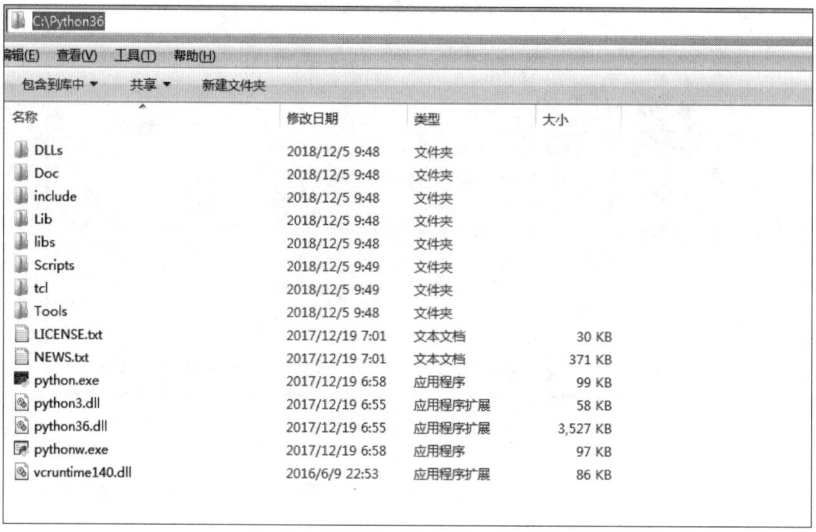

图 2.5　Python 目录

至此，Python 环境安装完成。

注意：本小节只介绍在 Windows 操作系统下安装 Python，对于其他操作系统下的安装，读者可参考其他资料。

2.1.2　搭建 Selenium 3 自动化环境

Selenium 自动化测试环境的搭建十分简单。本小节实战基于 Windows 操作系统来演示，如果是 Mac 或 Linux 操作系统，请自行参考其他搭建教程。

1. 在线安装 Selenium 环境

通过 Python 内置的 pip 命令来安装 Selenium。pip 可以理解为 Python 管家，负责安装和管理与 Python 相关的工具。直接安装最新版本的 Selenium 环境，打开 cmd 命令提示符界面，输入"pip install selenium"进行在线安装，示例如下：

```
C:\Users\23939>pip install selenium
Collecting selenium
  Downloading https://files.pythonhosted.org/packages/80/d6/4294f0b4bce4de0abf13e171
90289f9d0613b0a44e5dd6a7f5ca98459853/selenium-3.141.0-py2.py3-none-any.whl (904kB)
     100% |████████████████████████████████| 911kB 377kB/s
Requirement already satisfied: urllib3 in c:\python36\lib\site-packages (from
selenium) (1.23)
Installing collected packages: selenium
Successfully installed selenium-3.141.0
```

2. 离线安装 Selenium 环境

访问 Selenium 官网（https://www.seleniumhq.org/download），下载最新的 Selenium 安装包，解压后直接打开 setup.py 所在目录，执行 python setup.py install 命令，即可完成安装。

3. 验证 Selenium 环境

可以通过 pip 其他命令来查看 Selenium 的相关信息，如版本和安装位置等。使用命令 pip show selenium 查看 Selenium 版本信息，示例如下：

```
C:\Users\23939>pip show selenium
Name: selenium
Version: 3.141.0
Summary: Python bindings for Selenium
Home-page: https://github.com/SeleniumHQ/selenium/
Author: UNKNOWN
Author-email: UNKNOWN
License: Apache 2.0
Location: c:\python36\lib\site-packages
Requires: urllib3
Required-by: Appium-Python-Client
```

2.2 配置浏览器驱动

Selenium 调用浏览器做自动化测试，必须要有源生浏览器驱动的支持。本节介绍各种主流浏览器驱动的配置。

2.2.1 配置 IE 浏览器驱动

下载 IEDriver.exe，驱动下载地址为 http://selenium-release.storage.googleapis.com/index.html，根据操作系统版本选择对应的安装文件，下载后需要进行一些设置。

将 IEDriver.exe 文件所在的本地路径添加到系统环境变量中。IE 浏览器 "Internet 选项" 对话框的 "安全" 选项卡中有 4 个安全选项，分别是 Internet、本地 Intranet、受信任的站点和受限制的站点，每个安全选项均有"启用保护模式"复选框，全部选中，如图 2.6 所示。

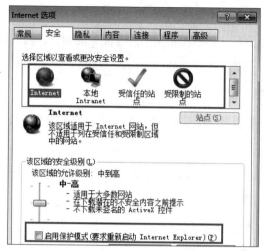

图 2.6 "Internet 选项"对话框

2.2.2 配置 Firefox 浏览器驱动

最新版本的 Selenium 3 不支持 47 版本以下的 Firefox 浏览器，所以应确保 Firefox 浏览器是最新版本。下载 Firefox 浏览器驱动，下载地址为 https://github.com/mozilla/geckodriver/releases，选择对应的操作系统版本下载。将 Firefox 浏览器驱动（geckodriver.exe）所在的文件路径追加到系统环境变量中，如图 2.7 所示。

图 2.7 Python 安装目录界面

将 Firefox 浏览器路径追加到系统环境变量中，如图 2.8 所示。

图 2.8　Firefox 安装目录界面

2.2.3　配置 Chrome 浏览器驱动

Chrome 浏览器驱动的下载地址为 http://chromedriver.storage.googleapis.com/index.html。下载驱动后，将谷歌浏览器驱动（chromedriver.exe）所在的文件路径追加到系统环境变量中，如图 2.9 和图 2.10 所示。

图 2.9　Python 安装目录界面

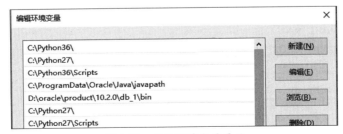

图 2.10　系统环境变量界面

注意：配置 Chrome 浏览器驱动时要注意浏览器和驱动的对应关系。

2.3 元素定位实战

本节主要结合众多自动化测试案例介绍元素定位实战,如单个元素定位、多个元素定位、By 类定位及引用 JavaScript、JQuery 辅助定位来加强元素定位实战技巧。在笔者看来,元素定位在自动化测试中非常重要。

2.3.1 单个元素定位实战

WebDriver 提供了多种内置定位方法,如 id、name、class、link_text、partial_link_text、tag、xpath 和 css 定位等。下面介绍一个百度首页案例。按 F12 键,会出现 Chrome 调试工具,单击调试工具上方的小箭头,定位到文本框位置上。在 HTML 页面下方显示的是页面文本框对应的 HTML 元素。拆分 HTML 元素,如图 2.11 所示。

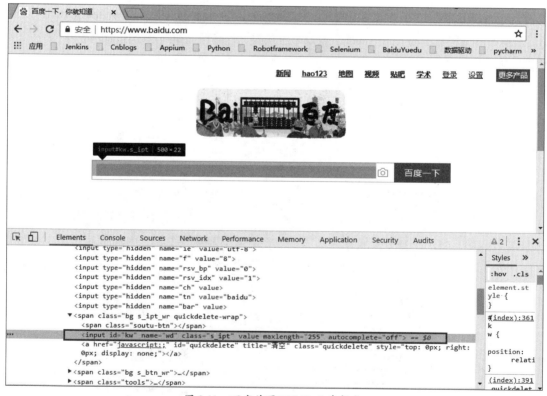

图 2.11 百度首页 HTML 元素拆分

在上面的案例中,百度文本框对应的元素是 <input id="kw" name="wd" class="s_ipt" value maxlength="255" autocomplete="off" >,其中 =(等号)左侧的 id、name、class、value maxlength 和

autocomplete 等都是元素的属性，就像人有身份证、姓名和身高等属性一样；等号右侧的 kw、wd、s_ipt、255、off 等都是属性值，就像人的身份证号、姓名和身高等。Selenium 工具就是通过识别这些属性来进行测试的。

1. 通过 find_element_by_id() 定位

示例如下：

```
from selenium import webdriver
driver = webdriver.Chrome()
driver.get('http://www.baidu.com')
driver.find_element_by_id("kw").send_keys('大道至简')
driver.quit()
```

2. 通过 find_element_by_name() 定位

示例如下：

```
from selenium import webdriver
driver = webdriver.Chrome()
driver.get('http://www.baidu.com')
driver.find_element_by_name("wd").send_keys('大道至简')
driver.quit()
```

3. 通过 find_element_by_class_name() 定位

示例如下：

```
from selenium import webdriver
driver = webdriver.Chrome()
driver.get('http://www.baidu.com')
driver.find_element_by_class_name("s_ipt").send_keys('大道至简')
driver.quit()
```

上述 3 个案例中，分别使用了 id、name、class_name 来定位百度文本框。send_keys() 是一个方法，表示向文本框中输入内容。在元素定位中，如果元素本身有 id、name 属性，则直接使用 id、name 定位，一般来说，id、name 属性值都是唯一的。相反，在使用 class 定位时，要检查 class_name 属性值在页面中是否唯一。如果不唯一，尽量不要用 class 去定位。举个简单的例子，一个班级里有两个叫张飞的同学，喊张飞时，这两个同学都会一起站起来，这样就会产生冲突。

通过浏览器自带的调试工具来验证属性的唯一性。按 F12 键，再按 Ctrl+F 组合键，搜索指定的元素属性值。如图 2.12 所示，class 属性值在整个页面显示数量只有 1 个，这时就可以使用 class 来定位了。有些情况下，class 属性值即使是唯一的，可能也会定位报错，这时就需要考虑其他定位方法。

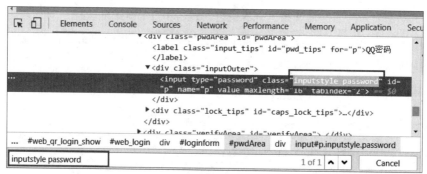

图 2.12 Chrome 调试工具界面

4. 通过 find_element_by_xpath() 定位

示例如下：

```
from selenium import webdriver
driver = webdriver.Chrome()
driver.get('http://www.baidu.com')
driver.find_element_by_xpath("//input[@id='kw']").send_keys(' 大道至简 ')
driver.quit()
```

// 表示相对路径，input 标签是文本框的所在标签，@id=kw 是文本框元素 id 的属性值。当然，也可以把 id 换成其他的定位，如 name 等。

示例如下：

```
from selenium import webdriver
driver = webdriver.Chrome()
driver.get('http://www.baidu.com')
driver.find_element_by_xpath("//form[@id='form']/span/input").send_keys(' 大道至简 ')
driver.quit()
```

//form[@id='form'] 表示父类元素，/ 表示下一级标签，/span/input 表示 span 标签下的 input 标签，如图 2.13 所示。

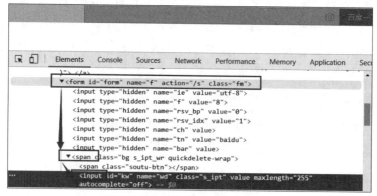

图 2.13 父子关系定位界面演示图

示例如下：

```
from selenium import webdriver
driver = webdriver.Chrome()
driver.get('http://www.baidu.com')
driver.find_element_by_xpath("//input[@id='kw' and @class='s_ipt']").send_keys('大道至简')
driver.quit()
```

//input 表示 HTML 页面相对路径下的 input 标签；@id='kw' and @class='s_ipt' 是 input 标签的两个属性，通过 id 和 class 两个属性加强元素的唯一性。

5. 通过 find_element_by_css_selector() 定位 1

示例如下：

```
from selenium import webdriver
driver = webdriver.Chrome()
driver.get('http://www.baidu.com')
driver.find_element_by_css_selector("input#kw").send_keys('大道至简')
driver.quit()
```

input#kw 表示 input 标签下 id 属性值是 kw 的元素。在 css 定位中，id 简写用"#"来表示，class 属性简写用"."来表示。

6. 通过 find_element_by_css_selector() 定位 2

示例如下：

```
from selenium import webdriver
driver = webdriver.Chrome()
driver.get('http://www.baidu.com')
driver.find_element_by_css_selector("form#form>span>input").send_keys('大道至简')
driver.quit()
```

form#form 表示 form 标签下 id 属性值是 form 的元素。在 css 定位中，层级关系用">"来表示。span>input 表示 span 标签下的 input 标签。有时为了简洁，也可以不写">"，直接写成 form#form span input。

2.3.2 多个元素定位实战

有时定位的元素属性值或标签名并不是唯一的，如一个页面有多个相同的 type 属性值或多个相同的标签名，如图 2.14 所示。

```html
<form class="form-horizontal">
  <div class="control-group">
    <label class="control-label" for="c1">selenium自动化测试</label>
    <div class="controls">
      <input type="checkbox" id="c1">
    </div>
  </div>
  <div class="control-group">
    <label class="control-label" for="c2">Robotframework自动化测试</label>
    <div class="controls">
      <input type="checkbox" id="c2">
    </div>
  </div>
  <div class="control-group">
    <label class="control-label" for="c3">Requests接口自动化测试</label> == $0
    <div class="controls">
      <input type="checkbox" id="c3">
    </div>
  </div>
</form>
```

图 2.14 多个元素属性界面

示例如下：

```python
from selenium import webdriver
driver = webdriver.Chrome()
driver.get('file:///' + os.path.abspath('checkbox.html'))
driver.find_elements_by_css_selector("input[type='checkbox']")[0].click()
driver.quit()
```

如图 2.14 所示，整个页面中有 3 个相同的 input 标签，且 input 标签下的 type 属性值 checkbox 都是一样的。这种情况可以使用 find_elements_by_css_seletor() 方法定位，该方法返回一个列表对象。通过 Python 索引下标来获取要定位的元素。[0] 表示第一个 input 标签下的 checkbox 属性值，[1] 表示第二个 input 标签下的 checkbox 属性值。

再看一个多个标签定位案例，如图 2.15 所示。

图 2.15 百度首页调试界面

示例如下：

```python
from selenium import webdriver
from time import sleep
driver = webdriver.Chrome()
driver.get("https://www.baidu.com")
tag = dr.find_elements_by_tag_name('input')
for t in tag:
if t.get_attribute('autocomplete') == 'off':
    t.send_keys('fighter007')
driver.find_element_by_id('su').click()
dr.quit()
```

如图 2.15 所示，当页面中有多个 input 标签时，可以通过标签本身的属性来定位。先获取所有 input 标签，然后通过 for 循环遍历每一个对象，最后判断获取的对象中 autocomplete 属性等于 off 的元素。

2.3.3　使用 By 类定位

使用 By 类定位，需要先导入 By 类。元素定位中的 id、name、xpath、class、link_text 和 css 分别在 By 类定位的语法是 By.ID、By.NAME、By.XPATH、By.CLASS_NAME、By.LINK_TEXT 和 By.CSS_SELECTOR，示例如下：

```python
from selenium import webdriver
from selenium.webdriver.common.by import By  # 导入By类
driver = webdriver.Chrome()
driver.get("https://www.baidu.com")
driver.find_element(By.ID,'kw').send_keys('大道至简')
driver.find_element(By.ID,'kw').click()
```

By 类定位的统一语法是 find_element(By.ID,'kw')，即通过 id 定位百度文本框，kw 是文本框的 id 属性值。By 类中的元素定位方法要注意区分大小写。

2.3.4　使用 JavaScript 定位

JavaScript 提供了多种元素定位方式，如 getElementsByClassName()、getElementByID()、getElementsByName()、getElementsByTagName() 和 document.querySelectorAll() 等。除 getElementByID() 定位返回单个 elements 元素定位外，其他的定位方式都是返回 list 对象，如图 2.16 所示。

图 2.16　简书登录界面

示例如下：

```
from selenium import webdriver
import time as t
dr = webdriver.Chrome()
dr.get('https://www.jianshu.com/sign_in')
t.sleep(2)
# 单击"注册"按钮
js_register = 'document.getElementById("js-sign-up-btn").click();'
dr.execute_script(js_register)
t.sleep(2)
```

使用 document.getElementById() 方法定位"注册"按钮，其中 execute_script() 方法用于调用 js 方法来执行 JavaScript 脚本。

```
# 单击"登录"按钮
js_class = 'document.getElementsByClassName("active")[0].click();'
dr.execute_script(js_class)
t.sleep(2)
```

使用 getElementsByClassName() 方法定位"登录"按钮，该方法返回一个 list 对象，其中 [0] 表示获取第一个 class 属性的元素。

```
js_input = 'document.getElementsByTagName("input")[2].value="username";'
dr.execute_script(js_input)
t.sleep(2)
```

使用 getElementsByTagName() 方法定位账号文本框，其中 [2] 表示获取整个页面的第三个 input 标签，value="username" 表示向文本框输入账号为 username。

```
# 使用 js 标签名定位输入密码框
js_passwd = 'document.getElementsByTagName("input")[3].value="password";'   # [3]索引值为 3
dr.execute_script(js_passwd)
t.sleep(2)
```

使用 document.getElementsByTagName() 方法定位密码文本框，其中 [3] 表示获取整个页面的第四个 input 标签，value="password" 表示向文本框输入密码为 password。

2.3.5 使用 JQuery 定位

JQuery 是 JavaScript 中的一个库，JQuery 中的 css 选择器经常会用于元素定位中的策略，示例如下：

```
# 定位登录链接
js_class = 'document.getElementsByClassName("active")[0].click();'
dr.execute_script(js_class)
t.sleep(2)
# 定位账号文本框，输入 "username"
js_input = 'document.getElementsByTagName("input")[2].value="username";' dr.execute_
script(js_input)
t.sleep(2)
# 定位密码文本框，输入 "password"
# js_passwd = 'document.getElementsByTagName("input")[3].value="password";'
dr.execute_script(js_passwd)
t.sleep(2)
# 使用 css 选择器定位 "登录" 按钮
css_btn = 'document.querySelectorAll(".sign-in-button")[0].click();'
dr.execute_script(css_btn)
```

JQuery 定位语法可以统一用 document.querySelectorAll() 来表示，该方法返回的是一个 list 对象，其中 .sign-in-button 表示 "登录" 按钮本身的 class 属性，[0] 表示返回第一个元素对象。

2.4 下拉框实战

在自动化测试中定位下拉框控件的场景很常见。WebDriver 提供了多种定位下拉框的方式，如引入 Select 类和元素二次定位等，本节进行下拉框实战。

2.4.1　value 属性定位

定位下拉框控件前，首先要引用 Select 类，该类提供了 3 种方式，即 value、text 和 index。下面介绍 value 方法的使用，如图 2.17 所示。

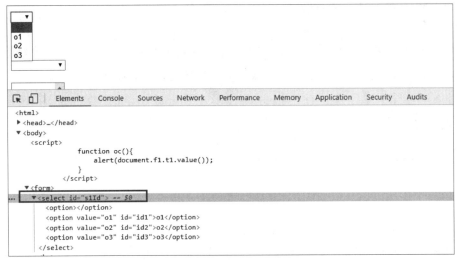

图 2.17　下拉框案例界面

示例如下：

```
from selenium import webdriver
from selenium.webdriver.common.by import By
from selenium.webdriver.support.ui import Select
from time import sleep
dr = webdriver.Chrome()
url = dr.get("file:///D:/xialakuang.html")
element = dr.find_element(By.ID,"s1Id")
Select(element).select_by_value("o1")
```

通过 id 定位到下拉框 Select 标签控件，将 element 对象传入 Select 类中，使用 Select 类下的 value 方法定位 value 值等于 o1 的元素，如图 2.18 所示。

图 2.18　下拉框 value 案例界面

2.4.2　index 属性定位

通过 Select 类下的 index 方法来定位下拉框中的值，示例如下：

```python
from selenium import webdriver
from selenium.webdriver.common.by import By          # 引入 By 类
from selenium.webdriver.support.ui import Select     # 引入 Select 类
from time import sleep
dr = webdriver.Chrome()
url = dr.get("file:///D:/xialakuang.html")
element = dr.find_element(By.ID,"s4Id")
Select(element).select_by_index(1)
```

select_by_index(1) 表示定位到第二个 option 标签，1 表示索引位为 1 的 option 标签。索引位默认从 0 开始。结果如图 2.19 所示。

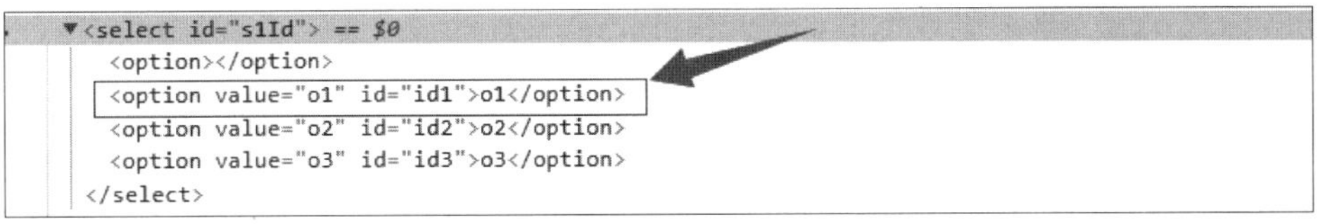

图 2.19　下拉框 index 案例界面

2.4.3　visible_text 属性定位

visible_text 文本属性一般是通过获取元素本身的文本描述来定位的，示例如下：

```python
from selenium import webdriver
from selenium.webdriver.common.by import By
from selenium.webdriver.support.ui import Select
from time import sleep
dr = webdriver.Chrome()
url = dr.get("file:///D:/xialakuang.html")
element = dr.find_element(By.ID,"s4Id")
Select(element).select_by_visible_text("o3")
```

select_by_visible_text("o3") 表示定位下拉框中 options 标签中文本描述为 o3 的元素。结果如图 2.20 所示。

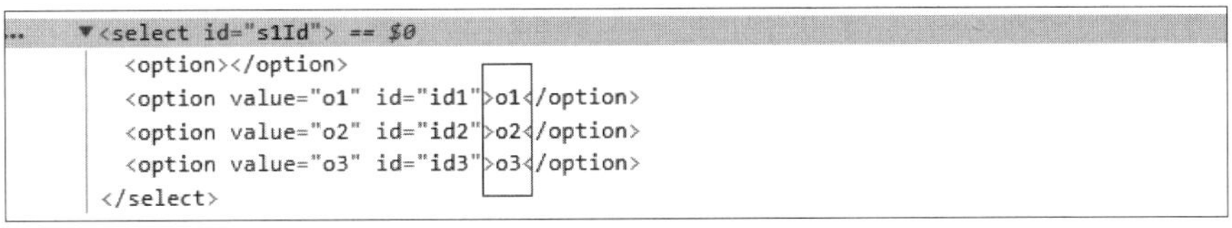

图 2.20　下拉框 visible_text 案例界面

2.4.4　元素二次定位实战

除 Select 类定位下拉框外，也可以通过二次定位操作下拉框。二次定位可以理解为将两步元素定位合并为一步实现，示例如下：

```
from selenium import webdriver
from time import sleep
dr = webdriver.Chrome()
url = dr.get("file:///D:/xialakuang.html")
dr.find_element_by_id("s4Id").find_element_by_xpath("//*[@id='id2']").click()
```

find_element_by_id("s4Id") 通过 id 定位到 Select 标签，find_element_by_xpath("//*[@id='id2']") 通过 xpath 方法定位 Select 标签下 id 属性值为 id2 的元素。

注意：使用二次定位方法操作下拉框内的元素时，必须增加 click 单击事件才会生效。

2.5 鼠标操作实战

WebDriver 提供了多种关于鼠标的操作方法，如模拟鼠标指针悬停、右击和双击等。本节进行鼠标操作实战。

2.5.1 鼠标指针悬停实战

本案例中使用 link_text 定位到百度首页的"设置"按钮。首先导入 ActionChains 类，再使用 move_to_element() 方法将鼠标指针悬停到某个元素上面。ActionChains() 方法需要驱动传入才生效，perform() 方法用于执行所有动作，示例如下：

```
from selenium import webdriver
from time import sleep
# 导入 ActionChains 类
from selenium.webdriver.common.action_chains import ActionChains
dr = webdriver.Chrome()
dr.get("https://www.baidu.com")
setting = dr.find_element_by_link_text("设置")
ActionChains(dr).move_to_element(setting).perform()
sleep(1)
dr.find_element_by_link_text("搜索设置").click()
sleep(5)
```

鼠标指针悬停到"设置"按钮后，弹出列表框，单击"搜索设置"选项，如图 2.21 所示。

图 2.21　鼠标指针悬停案例界面

2.5.2　鼠标右键实战

本案例中，首先将鼠标指针定位到要右击的元素上，然后通过 context_click() 方法执行鼠标右键操作，示例如下：

```python
# 导入 ActionChains 类
from selenium.webdriver.common.action_chains import ActionChains
dr = webdriver.Chrome()
dr.get("https://www.baidu.com")
sleep(3)
# 定位"百度一下"按钮元素
context = dr.find_element_by_id('su')
# 模拟鼠标右键操作
ActionChains(dr).context_click(context).perform()
sleep(5)
```

结果如图 2.22 所示。

图 2.22　鼠标右键案例界面

2.5.3 鼠标双击实战

本案例中,首先将鼠标指针定位到要双击的元素上,然后通过 double_click() 方法模拟鼠标双击操作,示例如下:

```python
from time import sleep
from selenium.webdriver.common.action_chains import ActionChains  # 导入 ActionChains 类
from selenium import webdriver
dr = webdriver.Chrome()
dr.get("https://www.baidu.com")
sleep(3)
dr.find_element_by_id('kw').send_keys('双击一下')
# 定位"百度一下"按钮元素
double = dr.find_element_by_id('su')
# 模拟鼠标双击操作
ActionChains(dr).double_click(double).perform()
sleep(5)
```

2.6 元素等待实战

在自动化测试任务中适当地引入元素等待可以降低元素定位的出错率,提高脚本稳定性。本节分别介绍元素定位过程中常见的 3 种等待,分别是强制等待、隐式等待和显示等待。

2.6.1 强制等待

强制等待就是 Python 语言 time 模块下的 sleep() 方法,它可以设置固定休眠时间。例如,sleep(5) 表示脚本执行到 sleep(5) 就强制等待 5s,等待 5s 过后再执行后面的语句,示例如下:

```python
from time import sleep
from selenium import webdriver
dr = webdriver.Chrome()
dr.get("https://www.baidu.com")
sleep(5)    # 强制等待 5s
dr.find_element_by_id('kw').send_keys('双击一下')
```

2.6.2 隐式等待

implicitly_wait() 是 WebDirver 提供的一个超时等待方法,比 sleep() 更加智能一些。隐式等待可以理解成在规定时间范围内,浏览器不停地刷新页面,直到找到目标元素。如果在规定时间内找不到目标元素,就抛出异常。不设置时默认是 0。

示例如下:

```python
from time import sleep
from selenium import webdriver
dr = webdriver.Chrome()
dr.get("https://www.baidu.com")
dr.implicitly_wait(30)    # 隐式等待 30s
dr.find_element_by_id('kw').send_keys('双击一下')
```

2.6.3 显示等待

显示等待可以理解为明确要等到某个元素的出现或是某个元素可单击。如果等不到,就一直等待下去;除非在规定时间内没找到该元素,那么就会抛出异常。

示例如下:

```python
from selenium import webdriver
from selenium.webdriver.common.by import By                         # 导入 By 类
from selenium.webdriver.support.ui import WebDriverWait             # 导入 WebDriverWait 类
from selenium.webdriver.support import expected_conditions as EC    # 导入 EC 模块
driver = webdriver.Chrome()
driver.get('https://mail.sina.com.cn/')
element = WebDriverWait(driver,5,0.5).until(EC.presence_of_element_located((By.ID,
        'freename')))
element.send_keys('hello')
driver.quit()
```

WebDriverWait() 是显示等待类。driver 是驱动;5 表示最长超时时间,单位为 s(秒);0.5 表示每隔 0.5,即检测元素是否存在的频率,单位为 s(秒);until 表示在等待期间,每隔一段时间调用这个传入的方法,直到返回值为 True;EC.presence_of_element_located() 方法表示只要有一个符合条件的元素加载出来就通过。

2.7 表单切换实战

在自动化测试过程中,有些情况下元素定位正常,但是脚本依然报错。这时很大可能是页面做了一些特殊处理,如页面中存在 iframe 标签等。本节进行表单切换案例实战。

2.7.1 单表单切换实战

单表单 iframe 的处理比较简单。如果 iframe 标签本身有可用的 id 或 name 属性，可以直接使用 switch_to.iframe() 方法去定位，如图 2.23 所示。

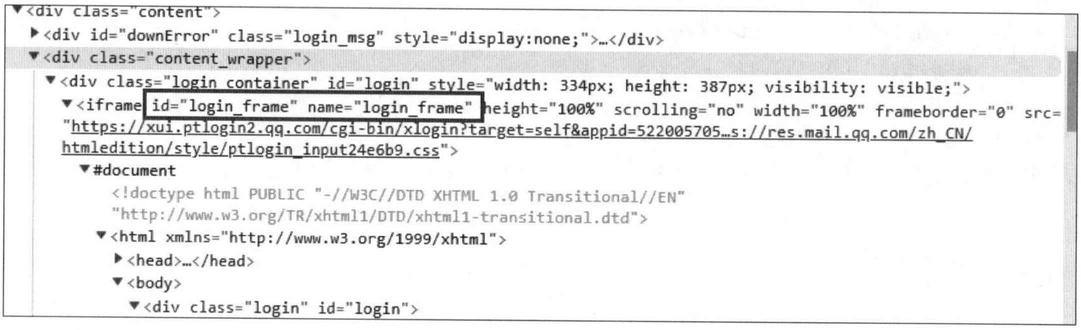

图 2.23 单表单 iframe 标签案例界面

示例如下：

```
from selenium import webdrvier
driver = webdriver.Chrome()
driver.get('https://mail.qq.com/cgi-bin/loginpage')
time.sleep(2)
driver.switch_to.frame('login_frame')              # 切换 iframe 标签
driver.find_element_by_name("email").send_keys('username')
driver.find_element_by_name("password").send_keys('password')
driver.find_element_by_id("login_button").click()
driver.switch_to.default_content()                 # 退出 iframe 标签
```

2.7.2 嵌套表单切换实战

在有些情况下，iframe 标签在整个页面中不止一个，如嵌套的 iframe 标签，如图 2.24 所示。

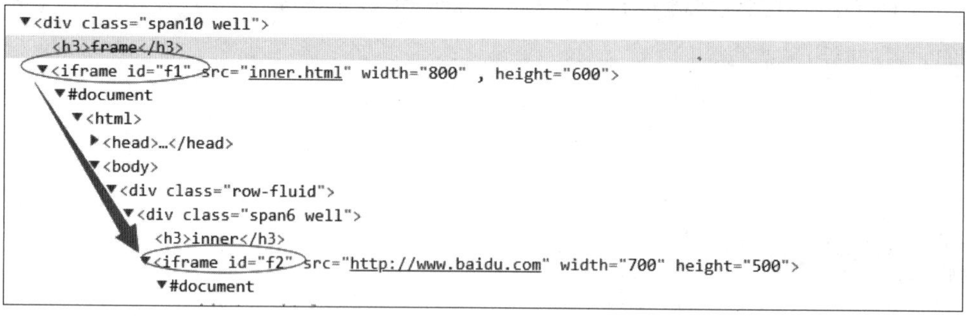

图 2.24 嵌套表单 iframe 标签案例界面

示例如下：

```
from selenium import webdrvier
driver = webdriver.Chrome()
driver.get('file:///E:/webdriver_api_demo/frame.html')
time.sleep(2)
# 先切换到最外层的 iframe 标签
driver.switch_to.frame('f1')
# 再切换到第二个 iframe 标签
driver.switch_to.frame('f2')
# 定位处在第二个 iframe 标签中的元素
driver.find_element_by_name("email").send_keys('username')
```

上述案例中，处理思路是：首先要切换到第一个 iframe 中，然后切换到第二个 iframe 标签，这样就可以正常定位到元素了。

2.7.3 平行表单切换实战

在有些情况下，iframe 标签在 HTML 页面中处于平行关系，如图 2.25 所示。

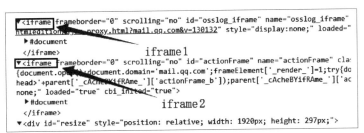

图 2.25 平行表单 iframe 标签案例界面

场景：假设当前处在 iframe1 标签中，现在要操作的元素在 iframe2 中。处理思路是：首先从 iframe1 标签中退出，然后切换到 iframe2 中去操作元素。

示例如下：

```
from selenium import webdrvier
driver = webdriver.Chrome()
driver.get('file:///E:/webdriver_api_demo/frame.html')
time.sleep(2)
# 默认在 iframe1 标签
driver.switch_to.frame('osslog_iframe')
# 退出 iframe1 标签
driver.switch_to.default_content()
# 切换到 iframe2 标签
driver.switch_to.frame('actionFrame')
# 操作 iframe2 标签下的元素
driver.find_element_by_name("email").send_keys('username')
```

注意：无论进入几层 iframe 标签，执行一次退出操作，即是最外层。

2.7.4 表单特殊情况处理

当 iframe 标签本身 id、name 属性值是一串动态变化的字符串或 iframe 本身没有可用的 id 或 name 属性时，可以借助层级定位来间接地定位 iframe 标签，如图 2.26 所示。

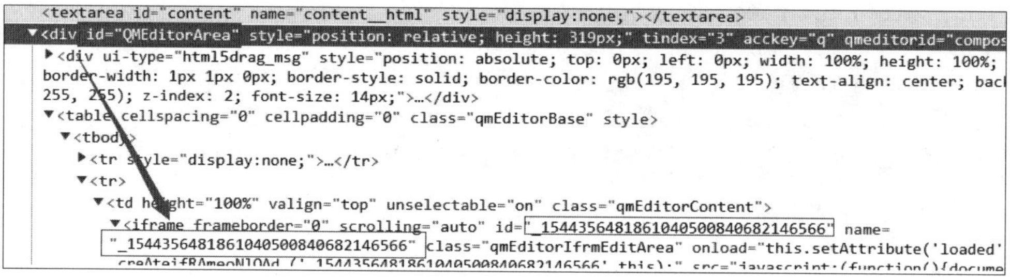

图 2.26 iframe 本身没有可用 id/name 属性界面

示例如下：

```
from selenium import webdrvier
driver = webdriver.Chrome()
driver.get('https://mail.qq.com/cgi-bin/loginpage')
time.sleep(2)
# 使用 xpath 层级定位 iframe 标签
Xpath = driver.find_element_by_xpath("//div[@id='QMEditorArea']/table/tbody/tr[2]/
        td/iframe")
driver.switch_to.frame(Xpath)    # 切换到 iframe 标签
driver.find_element_by_name("email").send_keys('username')
driver.find_element_by_name("password").send_keys('password')
driver.find_element_by_id("login_button").click()
```

第一步，通过 xpath 层级定位到 iframe 标签，将定位对象赋值给 Xpath；第二步，通过 swtich_to.frame() 方法切换到 iframe 标签。

2.8 窗口切换实战

做自动化测试时经常会遇到操作的元素不在当前窗口的场景。针对此种情况，本节进行多窗口切换实战案例演示。

2.8.1 GET 方法实战

当单击某个链接或按钮时，当前窗口会跳转到新的窗口。此时，可以使用 GET 方法获取新窗口的 URL，从而实现新窗口页面下的元素定位，如图 2.27 和图 2.28 所示。

图 2.27 窗口 1 界面

图 2.28 窗口 2 界面

示例如下：

```
from selenium import webdrvier
from time import sleep
driver = webdriver.Chrome()
driver.get('http://www.baidu.com')
driver.find_element_by_id("kw").send_keys("渗透吧")
driver.find_element_by_id("su").click()
# 第一个窗口下单击"渗透吧"链接
driver.find_element_xpath('//*[@id="1"]/h3/a').click()
# 使用 get() 方法获取跳转后的 URL 地址
driver.get('http://tieba.baidu.com/f?kw=%C9%F8%CD%B8&fr=ala0&tpl=5')
sleep(3)
# 操作跳转后所在窗口的页面元素
driver.find_element_by_link_text('进入贴吧').click()
```

2.8.2 SWITCH 方法实战

本案例中使用 window_handles 获取所有窗口，并返回一个 list 对象；然后通过 switch_to.window(window[index]) 方法来切换指定窗口，如 window(1) 表示切换到第二个窗口。

示例如下：

```
from selenium import webdriver
from time import sleep
from selenium.webdriver.common.by import By
dr = webdriver.Chrome()
url = "https://www.so.com/"
dr.get(url)
sleep(1)
dr.find_element_by_link_text("360 导航").click()
sleep(2)
# 获取所有窗口的句柄
windows = dr.window_handles
# 通过索引切换到第二个窗口
dr.switch_to.window(windows[1])
sleep(0.5)
# 在第二个窗口的文本框中输入 "第二个窗口"
dr.find_element_by_id("search-kw").send_keys(" 第二个窗口 ")
sleep(2)
# 切换到第一个窗口
dr.switch_to.window(windows[0])
# 在第一个窗口的文本框中输入 "第一个窗口"
dr.find_element_by_id("input").send_keys(" 第一个窗口 ")
```

2.9 警告框实战

WebDriver 提供了多种定位警告框的方法，如接收警告框、取消警告框和获取警告框的文本信息等。当单击某个操作触发警告框弹出时，需要对警告框进行确定或取消操作。本案例中，使用 switch_to.alert.text 方法获取警告框对应的文本信息，如图 2.29 所示。

图 2.29　警告框信息界面

示例如下：

```python
from selenium import webdriver
from time import sleep
# 鼠标指标悬停类
from selenium.webdriver.common.action_chains import ActionChains
from selenium.webdriver.support.select import Select
dr = webdriver.Chrome()
dr.get("https://www.baidu.com")
setting = dr.find_element_by_link_text("设置")
ActionChains(dr).move_to_element(setting).perform()
sleep(1)
dr.find_element_by_link_text("搜索设置").click()
sleep(5)
# 选择简体中文
dr.find_element_by_id("SL_1").click()
sleep(3)
# 下拉框的操作
select = dr.find_element_by_xpath("//select[@id='nr']")
Select(select).select_by_value("20")
sleep(3)
# 保存设置
dr.find_element_by_class_name("prefpanelgo").click()
# 输出警告信息
alert_text = dr.switch_to.alert.text
print(alert_text)
```

使用 switch_to.alert.accept() 方法接收警告框，此方法相当于单击警告框中的"确认"按钮操作，示例如下：

```python
from selenium import webdriver
from time import sleep
from selenium.webdriver.common.action_chains import ActionChains
from selenium.webdriver.support.select import Select
dr = webdriver.Chrome()
dr.get("https://www.baidu.com")
setting = dr.find_element_by_link_text("设置")              # 定位到"设置"
ActionChains(dr).move_to_element(setting).perform()          # 鼠标指针悬停设置选项
sleep(1)
dr.find_element_by_link_text("搜索设置").click()             # 单击"搜索设置"超链接
sleep(5)
dr.find_element_by_class_name("prefpanelgo").click()         # 单击保存设置
dr.switch_to.alert.accept()                                  # 确认警告框操作
```

使用 switch_to.alert.dismiss() 方法取消警告框，此方法相当于警告框中的"取消"按钮操作，示例如下：

```
from selenium import webdriver
from time import sleep
from selenium.webdriver.common.action_chains import ActionChains
from selenium.webdriver.support.select import Select
dr = webdriver.Chrome()
dr.get("https://www.baidu.com")
setting = dr.find_element_by_link_text(" 设置 ")
ActionChains(dr).move_to_element(setting).perform()
sleep(1)
dr.find_element_by_link_text(" 搜索设置 ").click()
sleep(5)
# 保存设置
dr.find_element_by_class_name("prefpanelgo").click()
# 拒绝警告框
dr.switch_to.alert.dismiss()
```

2.10 JavaScript 实战

本节演示 JavaScript 在自动化测试实战中几种比较常见的实战场景，如处理富文本、处理隐藏元素、处理 readonly 属性和处理浏览器滚动条等。

2.10.1 处理富文本实战

先来看一个处理富文本的案例，如图 2.30 所示。

图 2.30 富文本界面

示例如下：

```
from selenium import webdriver
from time import sleep
driver = webdriver.Chrome()
```

```
driver.get("http://localhost:8080")
# 定位富文本，并向富文本输入内容 A
js="document.getElementById('content_ifr').contentWindow.document.body.
innerHTML='%s'" %(A)
driver.execute_script(js)
```

document.getElementById() 通过 id 定位富文本，contentWindow.document.body.innerHTML 方法用于向富文本中写入内容。最后通过调用 js 来执行 JavaScript 脚本。

2.10.2　处理隐藏元素实战

在元素定位时，有时存在要定位的元素被隐藏了的情况。所谓元素被隐藏，指的是 style="display:none;" 这种情况，如图 2.31 所示。

```
▼<select id="s3" style="display:none;">
    <option value="-1">--请选择--</option>
    <option value="brtc1">python3</option>
    <option value="brtc2">selenium3</option>
    <option value="brtc3">po设计模式</option>
    <option value="brtc4">自动化持续集成</option>
</select>
```

图 2.31　隐藏元素界面

示例如下：

```
from selenium import webdriver
from selenium.webdriver.support.ui import Select
from selenium.webdriver.common.by import By
from time import sleep
dr = webdriver.Chrome()
dr.get('file:///E:/webdriver_api_demo/frame.html')
sleep(2)
# 设置元素为可见
js = 'document.querySelectorAll("select")[0].style.display="block";'
dr.execute_script(js)
sleep(3)
element = dr.find_element(By.ID,"s3")
Select(element).select_by_visible_text("po设计模式 ")
```

'document.querySelectorAll("select")[0].style.display="block";' 表示通过 JQuery 语法定位被隐藏的 Select 标签，style.display="block" 方法表示设置隐藏的元素为可见。

2.10.3　处理 readonly 属性实战

访问 12360 车票查询界面时，有时无法在出发日期文本框中输入日期，原因是元素本身有 readonly 属性存在。处理思路是：首先使用 JS 语法移除 readonly 属性，然后向文本框输入日期，如图 2.32 所示。

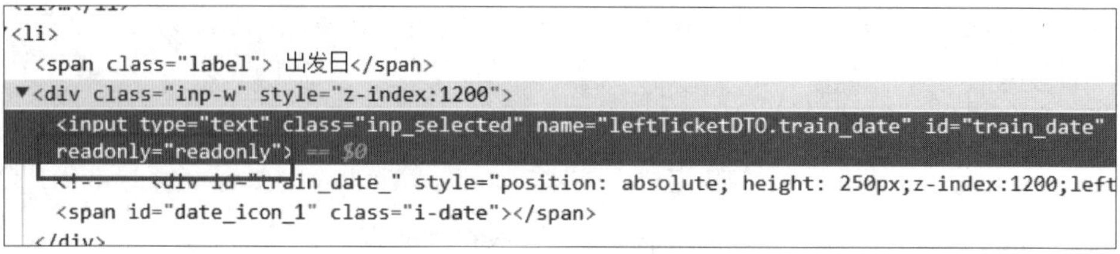

```
<li>
  <span class="label"> 出发日</span>
  <div class="inp-w" style="z-index:1200">
    <input type="text" class="inp_selected" name="leftTicketDTO.train_date" id="train_date"
    readonly="readonly"> == $0
    <!--   <div id="train_date_" style="position: absolute; height: 250px;z-index:1200;left
    <span id="date_icon_1" class="i-date"></span>
  </div>
```

图 2.32　readonly 属性界面（一）

移除 readonly 属性后，如图 2.33 所示。

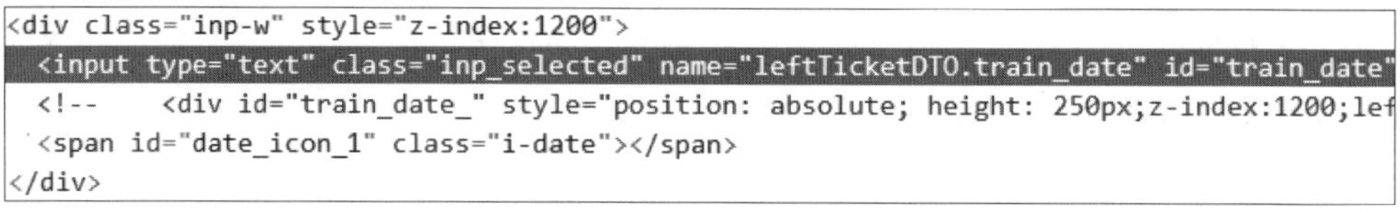

```
<div class="inp-w" style="z-index:1200">
  <input type="text" class="inp_selected" name="leftTicketDTO.train_date" id="train_date"
  <!--   <div id="train_date_" style="position: absolute; height: 250px;z-index:1200;lef
  <span id="date_icon_1" class="i-date"></span>
</div>
```

图 2.33　readonly 属性界面（二）

示例如下：

```python
from selenium import webdriver
from time import sleep
dr = webdriver.Chrome()
dr.get('https://kyfw.12306.cn/otn/leftTicket/init')
sleep(2)
# 移除 readonly 属性
js1 = "document.getElementById('train_date').removeAttribute('readonly');"
dr.execute_script(js1)
sleep(2)
dr.find_element_by_id('train_date').clear()   # 清空日期
dr.find_element_by_id('train_date').send_keys('2018-12-10')   # 输入最新日期
```

2.10.4　处理浏览器滚动条实战

如果被操作的元素超过了一定范围，则不能正常定位到元素，需要借助浏览器的滚动条来拖动屏幕，目的是让操作的元素出现在当前的屏幕上。针对这种场景，可以借助 JS 脚本来实现。WebDriver 提供了一个操作 JS 的方法 execute_script()，该方法可以直接定位到 JS 脚本，示例如下：

```python
from selenium import webdriver
from time import sleep
driver = webdriver.Chrome()
driver.get("http://www.baidu.com")                    # 访问百度搜索
driver.find_element_by_id("kw").send_keys("selenium")
driver.find_element_by_id("su").click()
sleep(3)
```

```
driver.execute_script("window.scrollTo(0,10000);")      # 将页面滚动条拖到底部
sleep(3)
driver.execute_script("window.scrollTo(10000,0);")      # 将滚动条移动到页面的顶部
sleep(3)
driver.quit()
```

window.scrollTo(0,10000) 表示以像素为单位，把滚动条滚动到 0 与 10000 位置处。其中 0 表示水平滚动条位置，10000 表示垂直滚动条位置。

除上述方法外，还可以引入键盘 Keys 类下的 DOWN、UP 方法来实现滚动条向下、向上滚动，示例如下：

```
from selenium import webdriver
from time import sleep
from selenium.webdriver.common.keys import Keys    # 引入键盘类
driver = webdriver.Chrome()
driver.maximize_window()
driver.get("http://www.baidu.com")                  # 访问百度搜索
driver.find_element_by_id("kw").send_keys("好奇心日报")
driver.find_element_by_id("su").click()
sleep(5)
# 将滚动条移动到页面底部
driver.find_element_by_xpath('//*[@id="page"]/a[10]').send_keys(Keys.DOWN)
sleep(5)
# 将滚动条移动到页面顶部
driver.find_element_by_xpath('//*[@id="s_tab"]/div/a[9]').send_keys(Keys.UP)
sleep(5)
driver.quit()
```

本案例中 ('//*[@id="page"]/a[10]') 表示百度搜索结果上方的"更多"按钮对应的元素，如图 2.34 所示。

图 2.34 "更多"按钮界面

'//*[@id="s_tab"]/div/a[9]' 表示百度搜索结果下方"下一页"按钮对应的元素，如图 2.35 所示。

图 2.35 "下一页"按钮界面

使用 Keys 类键盘实现滚动条拖动时，只需要指明滚动条滚动的目标位置，即所在屏幕上的任何一个元素就可以。

unittest 单元测试框架

unitttest 单元测试框架对于大部分 Python 爱好者来说并不陌生。unittest 单元测试框架不管在 Web 自动化测试还是接口自动化测试中,应用都是非常广泛的。它不仅可以组织、运行数量级的自动化测试用例,还提供丰富的断言方法、跳过测试及生成测试报告等功能。本节进行 unittest 单元测试框架的实战。

2.11.1 unittest 简介

unittest 是 Python 自带的一个单元测试框架。它不仅可以做单元测试,还适用于 Web 自动化测试用例的开发和执行。通过 unittest 官方文档可以了解到以下信息。

Python 单元测试框架有时称为 PyUnit,是 JUnit 的 Python 语言版本,由 Kent Beck 和 Erich Gamma 编写。JUnit 则是 Kent 的 Smalltalk 测试框架的 Java 版本。每一个都是它各自语言的事实上的标准单元测试框架。

unittest 的设计灵感最初来源于 JUnit 及其他语言中具有共同特征的单元框架。它支持自动化测试,在测试中使用 setup()(初始化)和 tearDown()(关闭销毁)操作,组织测试用例为套件(批量运行),以及把测试和报告独立开来。

下面来看一个 unittest 单元测试框架的基本案例,示例如下:

```python
import unittest
class TestStrSample(unittest.TestCase):
    def test_strendswich(self):
        self.assertEqual('foo'.endswith('o'),False)
    def test_split(self):
        s = 'my name is Fighter'
        self.assertEqual(s.split(),['my','name','is','Fighter'])
if __name__ == '__main__':
    unittest.main()
```

首先导入 unittest 单元测试框架,unittest 单元测试框架下的测试类都默认继承 unittest.TestCase 子类。测试类下的测试方法都必须以 test 开头,否则不会被执行。每一个测试方法中最后都使用了 assertEqual() 方法来判断实际结果是否等于预期结果。unittest.main() 方法提供了一个命令行接口,自动执行测试类下以 test 开头的测试方法。当执行完所有测试用例后,测试结果都会被记录并且写到测试报告中去。一般在测试执行结果中会有以下 3 种状态。

(1).(success):表示测试用例运行通过。

```
...
----------------------------------------------------------------------
Ran 3 tests in 0.000s
OK
```

（2）F(failed)：表示测试用例运行失败。

```
..F
======================================================================
FAIL: test_strendswich (__main__.TestStringMethods)
----------------------------------------------------------------------
Traceback (most recent call last):
  File "D:/project/One/webUI/webdriver_html/Pybase"
    self.assertEqual('foo'.endswith('o'),False)
AssertionError: True != False
```

（3）E(error)：表示测试脚本中存在错误。

```
..E
======================================================================
ERROR: test_strendswich (__main__.TestStringMethods)
----------------------------------------------------------------------
Traceback (most recent call last):
  File "D:/project/One/webUI/webdriver_html/Pybase.py",line 9,in test_strendswich
    self.assertEqual('foo'.endswth('o'),False)
AttributeError: 'str' object has no attribute 'endswth'
```

2.11.2 前置和后置

setUp() 和 tearDown() 方法可以在测试运行前后做一些操作。例如，Web 自动化测试过程中经常会把实例化浏览器、获取 URL、设置等待等操作存放在 setUp() 中，而 tearDown() 是测试完成后的清除工作，如关闭浏览器、数据库的还原等。

注意：setUp() 和 tearDown() 属于非必要条件，如果什么都不做，可以用 pass 来表示。

下面来看一个 setUp() 和 tearDown() 的案例，示例如下：

```
import unittest
class TestStringMethods(unittest.TestCase):
    def setUp(self):
        print('每条测试用例开始执行前做的操作.....')
    def test_isupper(self):
        self.assertTrue('FOO'.endswith('O'))
        self.assertFalse('Foo'.isupper())
        print('第一条测试用例')
    def test_strendswich(self):
```

```
                self.assertEqual('foo'.endswith('o'),True)
                print(' 第二条测试用例 ')
        def tearDown(self):
                print(' 每条测试用例执行完毕后做的操作 .....')
if __name__ == '__main__':
    unittest.main()
```

增加 setUp() 和 tearDown() 两部分测试固件后，运行结果如下：

```
..
每条测试用例开始执行前做的操作 .....
----------------------------------------------------------------------
第一条测试用例
Ran 2 tests in 0.000s
每条测试用例执行完毕后做的操作 .....
每条测试用例开始执行前做的操作 .....
OK
第二条测试用例
每条测试用例执行完毕后做的操作 .....
Process finished with exit code 0
```

测试用例执行顺序：setUp() → 第一条测试用例 → tearDown() → setUp() → 第二条测试用例 → tearDown()。也就是说，每执行一条测试用例，都要进行一次初始化操作和收尾工作。

2.11.3 常用断言方法

断言即在测试用例执行过程中，通过判断测试得到实际结果和预期结果是否相等。unittest.TestCase 类属性提供了多种断言方法，示例如下。

（1）assertEqual(a,b,[msg=' 测试失败时输出的信息 '])：断言 a 和 b 是否相等，相等则测试用例通过。

（2）assertNotEqual(a,b,[msg=' 测试失败时输出的信息 '])：断言 a 和 b 是否相等，不相等则测试用例通过。

（3）assertTrue(x,[msg=' 测试失败时输出的信息 '])：断言 x 是否为 True，是 True 则测试用例通过。

（4）assertFalse(x,[msg=' 测试失败时输出的信息 '])：断言 x 是否为 False，是 False 则测试用例通过。

（5）assertIs(a,b,[msg=' 测试失败时输出的信息 '])：断言 a 是否是 b，是则测试用例通过。

（6）assertNotIs(a,b,[msg=' 测试失败时输出的信息 '])：断言 a 是否是 b，不是则测试用例通过。

（7）assertIsNone(x,[msg=' 测试失败时输出的信息 '])：断言 x 是否为 None，是 None 则测试用例通过。

（8）assertIsNotNone(x,[msg=' 测试失败时输出的信息 '])：断言 x 是否为 None，不是 None 则

测试用例通过。

（9）assertIn(a,b,[msg=' 测试失败时输出的信息 ']): 断言 a 是否在 b 中，在 b 中则测试用例通过。

（10）assertNotIn(a,b,[msg=' 测试失败时输出的信息 ']): 断言 a 是否在 b 中，不在 b 中则测试用例通过。

（11）assertIsInstance(a,b,[msg=' 测试失败时输出的信息 ']): 断言 a 是 b 的一个实例，是则测试用例通过。

比较常用的断言方法有 assertEqual()、assertTrue()、assertIn()、assertIs() 等，其他方法可以作为备选。

测试固件整合 Web 自动化测试的示例如下：

```python
import unittest
from selenium import webdriver
from time import sleep
class TestWebUI(unittest.TestCase):
    def setUp(self):
        self.driver = webdriver.Chrome()
    def tearDown(self):
        self.driver.quit()
    def test_QQLogin(self):
        self.driver.get('https://mail.qq.com/cgi-bin/loginpage')
        self.assertEqual(self.driver.title,'登录QQ邮箱','页面跳转失败，请重新检查！')
    def test_MaoyanMovie(self):
        self.driver.get('https://maoyan.com/')
        self.assertEqual(self.driver.title,'猫眼电影 - 一网打尽好电影','页面跳转失败，
                        请重新检查！')
if __name__ == '__main__':
    unittest.main()
```

实例化浏览器、关闭浏览器分别定义在 setUp()、tearDown() 测试固件中。在 TestWebUI() 测试类下新增 test_QQLogin() 和 test_MaoyanMovie() 两条测试用例，每条测试用例用 title 获取验证信息，使用 assertEqual() 方法来判断实际结果和预期结果是否相等。如果不相等，就给出"页面跳转失败，请重新检查！"的提示信息。通过测试结果，可以看出运行两条测试用例消耗的时间为 17.197s。

运行结果如下：

```
..
----------------------------------------------------------------------
Ran 2 tests in 17.197s
OK
```

2.11.4　setUpClass() 和 tearDownClass()

使用 setUp() 和 tearDown() 方法在每次执行测试用例前都会先执行一次 setUp() 操作，然后执

行测试用例，最后执行 tearDown() 操作。假设有 100 条测试用例，如果把实例化浏览器定义在 setUp() 操作中，那么执行这 100 条用例就需要开启 100 次浏览器，这样会导致测试效率大大降低。

引入 setUpclass() 和 tearDownClass() 测试固件，可以保证运行所有测试用例时只需要开启一次浏览器和关闭一次浏览器即可完成测试任务，从而大大提高测试效率。

示例如下：

```
import unittest
from selenium import webdriver
from time import sleep
class TestWebUI(unittest.TestCase):
    @classmethod
    def setUpClass(cls):
        cls.driver = webdriver.Chrome()
    @classmethod
    def tearDownClass(cls):
        cls.driver.quit()
    def test_QQLogin(self):
        self.driver.get('https://mail.qq.com/cgi-bin/loginpage')
        self.assertEqual(self.driver.title,'登录QQ邮箱')
    def test_MaoyanMovie(self):
        self.driver.get('https://maoyan.com/')
        self.assertEqual(self.driver.title,'猫眼电影 - 一网打尽好电影')
if __name__ == '__main__':
    unittest.main()
```

案例中，@classmethod 装饰器结合 setUpClass() 和 tearDownUpClass() 方法完成启动一次浏览器和关闭一次浏览器动作。两条测试用例运行消耗的总时间长为 9.707s，从时间上看要比 setUp() 和 tearDown() 测试固件节省很多时间。

2.11.5　测试固件分离实战

编写 Web 自动化测试用例时会写很多个测试类，每个测试类下会有多个测试用例。但针对同一个测试项目而言，测试地址、实例化浏览器、关闭浏览器等操作都是固定的。可以把这部分测试固件单独分离出来，这样在测试类中的代码看起来会更加简洁、清晰，也提高了测试固件这部分代码的重用性。

针对 2.11.4 小节中的代码，新建 ...\Myunit.py 文件并做如下改进，示例如下：

```
import unittest
from selenium import webdriver
class TestWebUI(unittest.TestCase):
    @classmethod
    def setUpClass(cls):
```

```
        cls.driver = webdriver.Chrome()
    @classmethod
    def tearDownClass(cls):
        cls.driver.quit()
```

定义 TestWebUI() 测试类,把测试固件 setUpclass() 和 tearDownClass() 方法单独分离出来。修改 ...\testUnit.py 文件,示例如下:

```
from Myunit import *
class TestModle(TestWebUI):
    def test_QQLogin(self):
        self.driver.get('https://mail.qq.com/cgi-bin/loginpage')
        self.assertEqual(self.driver.title,'登录QQ邮箱 ')
    def test_MaoyanMovie(self):
        self.driver.get('https://maoyan.com/')
        self.assertEqual(self.driver.title,'猫眼电影 - 一网打尽好电影 ')
```

导入 Myunit.py 模块下的测试类和方法(*表示所有类和方法),其中 TestModle() 测试类继承 TestWebUI() 类,从而实例化浏览器。

2.11.6 生成 HTML 测试报告

unittest 单元测试框架可以将测试结果写到测试报告中,通过测试报告可以清晰地查看自动化测试用例的总数、通过数、失败数及失败原因等信息。

1. 配置 HTMLTestRunner.py 文件

将 HTMLTestRunner.py 文件复制到 Python 安装目录下的 Lib 目录,如 C:\Python36\Lib HTMLTestRunner.py。

2. 编写 allTest.py 文件,运行所有测试脚本

目录架构如下。

(1)...\TestCases:存放测试用例。

(2)...\TestCases__init__:声明 TestCases 包文件。

(3)...\TestCases\testUnit.py:自动化测试用例。

(4)...\TestCases\Myunit.py:存放测试固件。

(5)...\Reports:存放 HTML 测试报告。

(6)...\allTest.py:运行所有测试用例主程序文件。

编写主程序 ...\allTest.py 文件,代码如下:

```
import os,time,unittest
from HTMLTestRunner import HTMLTestRunner
```

```python
def getAllCases():
    ''' 获取 tesTcase 下的所有测试模块 '''
    Testsuite = unittest.defaultTestLoader.discover(
        start_dir = os.path.join(os.path.dirname(__file__),'TestCases'),
        pattern = 'test*.py')
    return Testsuite
def RunMain():
    ''' 生成测试报告，写入指定 Reports 目录 '''
    fp = open(os.path.join(os.path.dirname(__file__),'Reports', time.strftime("%Y_%m_%d_%H_%M_%S")+ 'report.html'),'wb')
    HTMLTestRunner(stream=fp,title='Python+Selenium 自动化测试实战 ',
                   description=' 基于 python 语言 UI 自动化测试 ').run(getAllCases())
if __name__ == '__main__':
    RunMain()
```

getAllCases() 方法下的 discover() 函数用于读取 TestCases 目录下以 test 开头的 .py 文件，并返回所有测试模块下的测试用例。RunMain() 方法用于生成测试报告，并将测试结果写入测试报告中。wb 模式用于读取二进制文件。time.strftime() 方法用于获取系统的当前时间，以便区分生成的不同测试报告名。

生成的 HTML 测试报告如图 2.36 所示。

图 2.36　生成的 HTML 测试报告

2.12　数据驱动测试实战

在自动化测试过程中，经常面临测试数据的维护和管理等问题。面对同一个功能如登录，登录流程每次都是一样的，只是测试数据不一样，这种情况在写测试用例时，需要创建多组不同的测试数据来检查登录功能的正确性。一种比较方便、高效的办法就是引用数据驱动测试。

2.12.1 DDT 简介与安装

DDT（Data-Driven Tests）数据驱动测试可以理解为：因测试数据的改变而驱动自动化测试的执行，最终引起测试结果的改变。通过数据驱动测试方法，每一组输入数据都对应一组测试用例，可以验证多组数据场景，如图 2.37 所示。

账号	密码	预期结果
23939899	admin	登录成功
23931192	admin	登录失败
23933349	admin	登录失败
39333299	admin	登录失败

图 2.37 DDT 测试数据界面

进行数据驱动测试前，需要安装 DDT 模块。打开 cmd 命令提示符界面，输入 "pip install ddt"，可以直接在线安装 DDT，示例如下：

```
C:\Users\23939>pip install ddt
Collecting ddt
  Using cached https://files.pythonhosted.org/packages/e4/d8/4649ee669e41760d824
63057215d93efa899ed0472a05b256b9ff7ba53c2/ddt-1.2.0-py2.py3-none-any.whl
Installing collected packages: ddt
Successfully installed ddt-1.2.0
```

2.12.2 DDT 在自动化测试中的应用

在进行数据驱动测试实战中，需要在测试类上使用 @ddt.ddt 装饰器，在测试用例上使用 @ddt.data 装饰器。@ddt.data 装饰器可以把参数作为测试数据，参数可以是单个值、列表、元组或字典。对于列表和元组，需要使用 @ddt.unpack 装饰器把元组和列表解析成多个参数。

示例如下：

```
import ddt
import unittest
Data = [{'name':"keep learing"},
        {'age':18},
        {'address':"深圳地区"}]
@ddt.ddt
class TestModules(unittest.TestCase):
    def setUp(self):
        print('testcase beaning....')
    def tearDown(self):
        print('testcase ending.....')
    @ddt.data(*Data)
```

```python
    def test_DataDriver(self,Data):
        print('DDT 数据驱动实战演示：',Data)
if __name__ == '__main__':
    unittest.main()
```

运行结果如下：

```
testcase beaning....
ddt 数据驱动实战演示：{'name': 'keep learing'}
testcase ending.....
testcase beaning....
ddt 数据驱动实战演示：{'age': 18}
testcase ending.....
testcase beaning....
ddt 数据驱动实战演示：{'address': ' 深圳地区 '}
testcase ending.....
Ran 3 tests in 0.005s
OK
```

上述案例中，不使用 @ddt.unpack 说明是两组测试数据，将 Data 内的每组数据分别作为参数传入 @ddt.data() 方法中，从而实现数据驱动测试。

使用 @ddt.unpack 装饰器解析列表或元组为多组参数，示例如下：

```python
import ddt
import unittest
Data = [['admin','admin',' 登录失败 '],
        ['admin','admin123',' 登录成功 '],
        ['','',' 登录失败 ']]
@ddt.ddt
class TestModules(unittest.TestCase):
    def setUp(self):
        print('testcase beaning....')
    def tearDown(self):
        print('testcase ending.....')
    @ddt.data(*Data)
    @ddt.unpack
    def test_DataDriver(self,user,passwd,text):
        print('ddt 数据驱动实战演示：',user)
        print('ddt 数据驱动实战演示：',passwd)
        print('ddt 数据驱动实战演示：',text)
if __name__ == '__main__':
    unittest.main()
```

运行结果如下：

```
testcase beaning....
ddt 数据驱动实战演示：admin
```

```
ddt 数据驱动实战演示：admin
ddt 数据驱动实战演示：登录失败
testcase ending.....
testcase beaning....
ddt 数据驱动实战演示：admin
ddt 数据驱动实战演示：admin123
ddt 数据驱动实战演示：登录成功
testcase ending.....
testcase beaning....
ddt 数据驱动实战演示：
ddt 数据驱动实战演示：
ddt 数据驱动实战演示：登录失败
testcase ending.....
```

Data 列表内存储的 3 组测试数据分别是账号、密码和文本信息。使用 @ddt.upack 方法将 Data 列表内的数据进行分解，分解后 user、passwd 和 text 分别对应每组测试数据中的账号、密码和文本信息。

DDT 在自动化测试中的应用示例如下：

```
import ddt,unittest
from time import sleep
from selenium import webdriver
from selenium.webdriver.common.by import By
from selenium.webdriver.support.ui import WebDriverWait        # 导入 WebDriverWait 类
from selenium.webdriver.support import expected_conditions as EC    # 导入 EC 模块
def readData():
    return [
            ['','',' 请输入账号密码 '],
            ['admin123@sohu.com','',' 请输入账号密码 '],
            ['admin111@sohu.com','',' 请输入账号密码 '],
            ['','a123456789',' 请输入账号密码 '],
            ]
@ddt.ddt
class TestLogin(unittest.TestCase):
    def setUp(self):
        self.driver = webdriver.Chrome()
        self.testUrl = "https://mail.sohu.com/fe/#/login"
    def tearDown(self):
        self.driver.quit()
    def by_css(self,usernameloc):
        ''' 重写 css 定位 '''
        return self.driver.find_element_by_css_selector(usernameloc)
    def getassertText(self):
        ''' 获取验证信息 '''
        try:
            sleep(2)
            loctor = (By.CSS_SELECTOR,'.tipHolder.ng-binding')
```

```
            WebDriverWait(self.driver,5,0.5).until(EC.presence_of_element_located
((loctor)))
            return self.by_css('.tipHolder.ng-binding').text
        except Exception as message:
            print('元素定位报错！报错原因是：{}'.format(message))
    @ddt.data(*readData())
    @ddt.unpack
    def test_souhuLogin(self,user,passwd,text):
        self.driver.get(self.testUrl)
        sleep(3)
        self.by_css('.addFocus.ipt-account.ng-pristine.ng-valid').send_keys(user)
        self.by_css('.addFocus.ng-pristine.ng-valid').send_keys(passwd)
        self.by_css('.btn-login.fontFamily').click()
        self.assertEqual(self.getassertText(),text)
if __name__ == '__main__':
    unittest.main()
```

readData() 方法中定义了 3 组测试数据，分别是登录账号、密码和验证信息。测试用例上的 @ddt.data 把 readData() 方法中的参数作为测试数据，@ddt.unpack 用于将 readData() 方法中的列表分解为多组测试数据。by_css() 方法是对源生 css 定位的二次封装，getassertText() 方法用来获取页面的文本信息以做断言使用。

2.12.3　Excel 自动化测试实战

在实际自动化测试任务中，很多情况下会将测试脚本中的测试数据和测试用例分开管理，一般会将测试数据单独存放在 Excel 中进行维护。Python 提供了很多第三方的模块来读取 Excel 数据，如 xlrd、xlwt 和 xlutils 等。本小节主要讨论 xlrd 在自动化测试中的应用。

首先安装 xlrd 模块，打开 cmd 命令提示符界面，输入"pip install xlrd"进行在线安装，示例如下：

```
C:\Users\23939>pip install xlrd
Collecting xlrd
Using cached https://files.pythonhosted.org/packages/07/ec4e2xlrd-1.1.0-py2.py3-
none-any.whl
Installing collected packages: xlrd
Successfully installed xlrd-1.1.0
```

出现上述结果表示安装成功。

示例如下：

```
import xlrd
import ddt,unittest
from time import sleep
from selenium import webdriver
from selenium.webdriver.common.by import By
```

```python
from selenium.webdriver.support.ui import WebDriverWait      # 导入 WebDriverWait 类
from selenium.webdriver.support import expected_conditions as EC   # 导入 EC 模块
def readUserName(row):
    '''读取用户名'''
    book = xlrd.open_workbook('datainfo.xlsx','r')
    table = book.sheet_by_index(0)
    return table.row_values(row)[0]
def readPasswd(row):
    '''读取用户名'''
    book = xlrd.open_workbook('datainfo.xlsx','r')
    table = book.sheet_by_index(0)
    return table.row_values(row)[1]
def readAssertText(row):
    '''读取预期结果'''
    book = xlrd.open_workbook('datainfo.xlsx','r')
    table = book.sheet_by_index(0)
    return table.row_values(row)[2]
```

readUserName()、readPasswd()、AssertText() 方法分别用于读取 datainfo.xlsx 表中每一行的用户名、密码和文本验证信息。

```python
class TestSouHuLogin(unittest.TestCase):
    def setUp(self):
        self.driver = webdriver.Chrome()
        self.testUrl = "https://mail.sohu.com/fe/#/login"
    def tearDown(self):
        self.driver.quit()
    def by_css(self,usernameloc):
        '''重写 css 定位'''
        return self.driver.find_element_by_css_selector(usernameloc)
    def getassertText(self):
        '''获取验证信息'''
        try:
            sleep(2)
            loctor = (By.CSS_SELECTOR,'.tipHolder.ng-binding')
            WebDriverWait(self.driver,5,0.5).until(EC.presence_of_element_located((loctor)))
            return self.by_css('.tipHolder.ng-binding').text
        except Exception as message:
            print('元素定位报错!报错原因是:{}'.format(message))
    def souhuLogin(self,user,passwd):
        '''封装登录功能'''
        self.by_css('.addFocus.ipt-account.ng-pristine.ng-valid').send_keys(user)
        self.by_css('.addFocus.ng-pristine.ng-valid').send_keys(passwd)
        self.by_css('.btn-login.fontFamily').click()
    def test_souHuLogin_001(self):
```

```
            '''账号和密码为空：登录失败'''
            self.driver.get(self.testUrl)
            sleep(3)
            self.souhuLogin(readUserName(1),readPasswd(1))
            self.assertEqual(self.getassertText(),readAssertText(1))
    def test_souHuLogin_002(self):
            '''账号正确和密码为空：登录失败'''
            self.driver.get(self.testUrl)
            self.souhuLogin(readUserName(2),readPasswd(2))
            self.assertEqual(self.getassertText(),readAssertText(2))
    def test_souHuLogin_003(self):
            '''账号错误和密码为空：登录失败'''
            self.driver.get(self.testUrl)
            self.souhuLogin(readUserName(3),readPasswd(3))
            self.assertEqual(self.getassertText(),readAssertText(3))
    def test_souHuLogin_004(self):
            '''账号为空和密码正确：登录失败'''
            self.driver.get(self.testUrl)
            self.souhuLogin(readUserName(4),readPasswd(4))
            self.assertEqual(self.getassertText(),readAssertText(4))
if __name__ == '__main__':
    unittest.main()
```

在 TestSouHuLogin 测试类中，对 css 定位并进行二次封装为 by_css() 方法，将搜狐邮箱登录功能封装为 souhuLogin() 方法，调用 souhuLogin() 方法时只需要输入参数 user、passwd 即可。此外，还分别定义了 4 组测试用例进行登录功能的验证。getassertText() 方法用于获取页面实际验证信息，并与预期结果 readAssertText() 方法返回的文本值进行比对，如果信息一致则测试用例通过，反之，则测试用例失败。

2.12.4 Excel 整合 DDT 自动化测试实战

首先在 Excel 表中写入以下数据，如图 2.38 所示。

账号	密码	预期结果
		请输入账号密码
admin111@sohu.com		请输入账号密码
admin222@sohu.com		请输入账号密码
	a1234567	请输入账号密码

图 2.38 DDT+Excel 测试数据界面

示例如下：

```
import xlrd
import ddt,unittest
from time import sleep
```

```python
from selenium import webdriver
from selenium.webdriver.common.by import By
from selenium.webdriver.support.ui import WebDriverWait      # 导入 WebDriverWait 类
from selenium.webdriver.support import expected_conditions as EC    # 导入 EC 模块
def readData():
    book = xlrd.open_workbook('datainfo.xlsx','r')  # 读取 datainfo.xlsx 表
    table = book.sheet_by_index(0)                   # 获取第一个 sheet
    newRows = []
    for rowValue in range(1,table.nrows):
        newRows.append(table.row_values(rowValue,0,table.ncols))
    return newRows              # 返回新的 newRows
```

readData() 方法用于读取 datainfo.xlsx 表中每一行的用户名、密码和文本信息,然后将读取结果分别依次追加到 newRows。

```python
@ddt.ddt
class TestLogin(unittest.TestCase):
    def setUp(self):
        self.driver = webdriver.Chrome()
        self.testUrl = "https://mail.sohu.com/fe/#/login"
    def tearDown(self):
        self.driver.quit()
    def by_css(self,usernameloc):
        ''' 重写 css 定位 '''
        return self.driver.find_element_by_css_selector(usernameloc)
    def getassertText(self):
        ''' 获取验证信息 '''
        try:
            sleep(2)
            loctor = (By.CSS_SELECTOR,'.tipHolder.ng-binding')
            WebDriverWait(self.driver,5,0.5).until(EC.presence_of_element_located((loctor)))
            return self.by_css('.tipHolder.ng-binding').text
        except Exception as message:
            print('元素定位报错!报错原因是:{}'.format(message))
    @ddt.data(*readData())
    @ddt.unpack
    def test_souhuLogin(self,user,passwd,text):
        self.driver.get(self.testUrl)
        sleep(3)
        self.by_css('.addFocus.ipt-account.ng-pristine.ng-valid').send_keys(user)
        self.by_css('.addFocus.ng-pristine.ng-valid').send_keys(passwd)
        self.by_css('.btn-login.fontFamily').click()
        self.assertEqual(self.getassertText(),text)
if __name__ == '__main__':
    unittest.main()
```

引入 ddt 模块，在测试类上增加 @ddt.ddt 装饰器，在测试用例上增加 @ddt.data、@ddt.unpack 方法。其中 @ddt.data 方法用于把 newRows 参数作为测试数据，newRows 参数实际是一个嵌套的列表；@ddt.unpack 方法用于把测试数据分解为多个值并且作为实际参数传入测试用例中的 user、passwd 和 text 中。getassertText() 方法用来获取页面实际文本信息，然后与 text 预期文本进行比较，验证信息一致则通过。

2.12.5 YAML 自动化测试实战

YAML 是一种直观的能够被计算机识别的数据序列化格式，容易阅读，并且容易和脚本语言交互。YAML 类似于 XML，但是语法比 XML 简单得多；而对于 JSON，YAML 可以写成规范化的配置文件。此外，不管是做 Web 自动化测试还是接口自动化测试，都可以使用 YAML 来管理测试数据，这种方法也比较简单高效。

YAML 的安装非常简单，可以直接在线使用 pip 命令进行安装。打开 cmd 命令提示符界面，输入 "pip install pyyaml" 进行在线安装，示例如下：

```
C:\Users\23939>pip install pyyaml
Collecting pyyaml
  Downloading https://files.pythonhosted.org/packages/4f/ca/5fad249c5032270540c24d21
89b0ddf1396aac49b0bdc548162edcf14131/PyYAML-3.13-cp36-cp36m-win_amd64.whl (206kB)
    100% |████████████████████████████████| 215kB 98kB/s
Installing collected packages: pyyaml
Successfully installed pyyaml-3.13
```

进入 Python 交互模式，来验证 YAML 是否安装成功，如下所示表示安装成功：

```
C:\Users\23939>python
Python 3.6.4 (v3.6.4:d48eceb,Dec 19 2017,06:54:40) [MSC v.1900 64 bit (AMD64)] on win32
Type "help","copyright","credits" or "license" for more information.
>>> import yaml
>>>
```

YAML 在自动化测试中使用 data.yaml 文件存储的数据信息如下，同时注意 YAML 文件的扩展名必须是以 .yaml 结尾。

```
userNull:
  username: ""
  password: ""
  assertText: " 请输入账号密码 "
passNull:
  username1: "amdin888"
  password1: ""
  assertText1: " 请输入账号密码 "
```

示例如下：

```python
import unittest,yaml
from time import sleep
from selenium import webdriver
def readYaml():
    ''' 获取所有 YAML 数据 '''
    f = open('data.yaml','r',encoding='utf-8')
    data = yaml.load(f)
    f.close()
    return data
```

导入 YAML 模块，readYaml() 方法下的 open() 方法用于打开 data.yaml 文件，为了防止乱码可使用 encoding="utf-8" 来声明。yaml.load() 方法用于读取 data.yaml 文件下的所有数据。close() 方法用于关闭整个 yaml 文件，如果不关闭会产生 ResourceWarning 提示，示例如下：

```python
class TestLogin(unittest.TestCase):
    def setUp(self):
        self.driver = webdriver.Chrome()
        self.testUrl = "https://mail.sohu.com/fe/#/login"
    def tearDown(self):
        self.driver.quit()
    def by_css(self,usernameloc):
        ''' 重写 css 定位 '''
        return self.driver.find_element_by_css_selector(usernameloc)
    def getassertText(self):
        ''' 获取验证信息 '''
        try:
            return self.by_css('.tipHolder.ng-binding').text
        except Exception as message:
            print('元素定位报错！报错原因是：{}'.format(message))
    def souhuLogin(self,user,passwd):
        ''' 封装登录功能 '''
        self.by_css('.addFocus.ipt-account.ng-pristine.ng-valid').send_keys(user)
        self.by_css('.addFocus.ng-pristine.ng-valid').send_keys(passwd)
        self.by_css('.btn-login.fontFamily').click()
    def test_souHuLogin_001(self):
        ''' 账号正确和密码为空：登录失败 '''
        self.driver.get(self.testUrl)
        sleep(3)
        self.souhuLogin(readYaml()['userNull']['username'],readYaml()['userNull']['password'])
        self.assertEqual(self.getassertText(),readYaml()['userNull']['assertText'])
    def test_souHuLogin_002(self):
        ''' 账号错误和密码为空：登录失败 '''
        self.driver.get(self.testUrl)
```

```
        sleep(3)
        self.souhuLogin(readYaml()['passNull']['username1'],readYaml()['passNull']
['password1'])
        self.assertEqual(self.getassertText(),readYaml()['passNull']['assertText1'])
if __name__ == '__main__':
    unittest.main()
```

readYaml() 方法返回所有的 YAML 数据，然后通过索引来获取指定的测试数据。例如，['username']、['password'] 和 'assertText' 分别访问 userNull 节点下的用户名信息、密码信息和预期文本信息。

2.12.6 parameterized 参数化实战

parameterized 翻译过来是参数化，是对 unittest 单元测试框架的参数化扩展。简单来说，它可以结合 unittest 单元测试框架进行一些参数化策略的应用。

1. 在线安装 parameterized

parameterized 的安装十分简单，可以直接在线安装。打开 cmd 命令提示符界面，输入 "pip install parameterized" 进行在线安装。如下所示表示安装成功：

```
C:\Users\23939>pip3 install parameterized
Collecting parameterized
  Downloading https://files.pythonhosted.org/packages/65/d4/b0b626eb263a4c2aa3ca3cd2
0ea3db410db837f7f6b5d3fc4a6c4bee3631/parameterized-0.6.1-py2.py3-none-any.whl
Installing collected packages: parameterized
Successfully installed parameterized-0.6.1
C:\Users\23939>
```

2. parameterized 在自动化测试中的应用

示例如下：

```
import unittest
from selenium import webdriver
from time import sleep
from parameterized import parameterized    # 导入参数化模块
class LoginTest(unittest.TestCase):
    @classmethod
    def setUpClass(cls):
        cls.driver = webdriver.Chrome()
        cls.testUrl = 'https://mail.sohu.com/fe/#/login'
    @classmethod
    def tearDownClass(cls):
        cls.driver.quit()
```

```python
    def by_css(self,usernameloc):
        '''重写css定位'''
        return self.driver.find_element_by_css_selector(usernameloc)
    def getassertText(self):
        '''获取验证信息'''
        try:
            return self.by_css('.tipHolder.ng-binding').text
        except Exception as message:
            print('元素定位报错！报错原因是：{}'.format(message))
    def souhuLogin(self,user,passwd):
        '''封装登录功能'''
        self.by_css('.addFocus.ipt-account.ng-pristine.ng-valid').send_keys(user)
        self.by_css('.addFocus.ng-pristine.ng-valid').send_keys(passwd)
        self.by_css('.btn-login.fontFamily').click()
```

上述代码对搜狐邮箱登录、css 元素定位和登录后的验证信息功能进行二次封装，封装后的方法分别是 souhuLogin()、by_css() 和 getassertText()。

```python
    @parameterized.expand([
                            ('','','请输入账号密码'),
                            ('admin111@sohu.com','','请输入账号密码'),
                            ('','a123456789','请输入账号密码')])
    def test_login(self,username,password,assert_text):
        # 登录系统
        self.driver.get(self.testUrl)
        sleep(3)
        self.souhuLogin(username,password)
        self.assertEqual(self.getassertText(),assert_text)
if __name__ == '__main__':
    unittest.main(verbosity=2)
```

@parameterized.expand() 方法用来存放多组测试数据，每组测试数据都会作为实参传递给测试用例中的 username、password 和 assert_text。getassertText() 方法返回的登录页面实际结果和 assert_text 定义好的预期结果进行比较，从而实现参数化登录过程。

2.13 发送邮件实战

SMTP 是一种简单的邮件传输协议、Python 默认支持 SMTP。使用 SMTP 可以构造纯文本的邮件和带附件的邮件。SMTP 提供 smtplib 和 email 两个主要模块，其中 smtplib 负责发送邮件，email 负责构造邮件。本节进行 SMTP 邮件发送实战。

2.13.1 纯文本的邮件实战

以 QQ 邮箱为例,首先设置 QQ 邮箱授权码。进入邮箱,单击"设置"→"账户"选项,开启 POP3/SMTP 服务。默认是关闭状态,单击"开启"按钮即可,如图 2.39 所示。

图 2.39　POP3/SMTP 服务设置界面

开启后,需要配置邮件客户端。编辑短信"配置邮件客户端"到号码"1069 0700 69",如图 2.40 所示。

图 2.40　配置邮件客户端界面

发送信息完成,单击"我已发送"按钮,会出现如图 2.41 所示的界面。

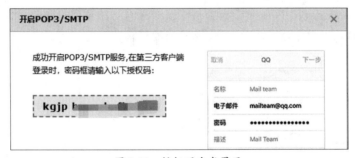

图 2.41　授权码生成界面

示例如下:

```
import smtplib                               # 调用 SMTP 发件服务
from email.mime.text import MIMEText         # 导入做纯文本的邮件模板类
```

```
smtpsever = 'smtp.qq.com'                    # QQ 邮箱服务器
sender = 'qq 邮箱账号 @qq.com'                # 发送者邮箱
psw = "hcygozfxeassddhhb"                    # 配置邮箱客户端生成的 QQ 邮箱授权码
receiver = '126 邮箱账号 @126.com'            # 接收者邮箱
port = 465                                   # QQ 邮箱服务器默认端口号
```

配置发送参数，如连接邮箱服务器，授权码，定义接收者、发送者邮箱账号和端口等。

```
msg = MIMEText(body,'html','utf-8')          # 邮件正文内容
msg['from'] = qq 邮箱账号 @qq.com'            # 发送者账号
msg['to'] = '126 邮箱账号 @qq.com'            # 接收者账号
msg['subject'] = " 这个是纯文本发送的邮件示例 "
```

编写正文内容到邮件中。

```
smtp = smtplib.SMTP_SSL(smtpsever,port)      # 调用发件服务
smtp.login(sender,psw)                       # 通过发送者的邮箱账号和授权码登录邮箱
smtp.sendmail(sender,receiver,msg.as_string())   # 发送邮件，信息以字符串方式保存
smtp.quit()                                  # 关闭邮件服务
```

QQ 邮箱的 SMTP 服务在登录时使用授权码登录，登录采用 SSL 方式，最后关闭服务。邮件发送界面如图 2.42 所示。

图 2.42　邮件发送界面

2.13.2　带附件的邮件实战

以 QQ 邮箱为例，向 126 邮箱发送带附件的邮件，这里需要用到 MIMEMultipart 类，示例如下：

```
import smtplib
from email.mime.text import MIMEText                        # 导入做纯文本的邮件模板类
from email.mime.multipart import MIMEMultipart              # 导入 MIMEMultipart 类
# 发邮件相关参数
smtpsever = 'smtp.qq.com'                    # QQ 邮箱服务器
sender = '239xxxxx@qq.com'                   # 发送者邮箱
psw = "xxxxxxxxxxxxxxxx"                     # QQ 邮箱授权码
receiver = 'xxxxx@126.com'                   # 接收者邮箱账号
port = 465                                   # QQ 邮箱服务器默认端口号
```

定义发送邮件的相关参数设置。

```
filepath = r"./readme.txt"                    # 编辑邮件的内容
with open(filepath,'rb') as fp:               # 读文件
    mail_body=fp.read()
```

open() 方法用于打开 readme.txt，并以 read() 方法读取所有内容。

```
# 主题
msg = MIMEMultipart()
msg["from"] = sender
msg["to"] = receiver
msg["subject"] = " 带附件的邮件发送模板主题 "
```

MIMEMultipart() 方法表示构造的邮件由多个部分组成。

```
body = MIMEText(mail_body,"html","utf-8")
msg.attach(body)
att = MIMEText(mail_body,"base64","utf-8")
att["Content-Type"] = "application/octet-stream"
att["Content-Disposition"] = 'attachment;filename="test_report.html"'
msg.attach(att)
```

application/octet-stream 表明返回的是一个二进制的文件，客户端收到这个声明后，会根据文件扩展名来判断。test_report.html 是以 .html 结尾的文件名。

```
try:
    smtp = smtplib.SMTP()
    smtp.connect(smtpsever)                   # 连接 QQ 邮箱服务器
    smtp.login(sender,psw)                    # 调用发件服务
except:
    smtp = smtplib.SMTP_SSL(smtpsever,port)
    smtp.login(sender,psw)                    # 登录邮箱
smtp.sendmail(sender,receiver,msg.as_string())  # 发送邮件
smtp.quit()
```

查看生成的带附件的邮件，如图 2.43 所示。

图 2.43　带附件的邮件界面

2.13.3 yagmail 发送邮件实战

yagmail 发送邮件相对 SMTP 服务更为简单和便捷。yagmail 是 Python 的一个第三方库。使用前先安装 yagmail 模块，使用在线安装方式，通过 pip 命令安装。打开 cmd 命令提示符界面，输入"pip install yagmail"，如下所示表示安装成功：

```
C:\Users\23939>pip install yagmail
Collecting yagmail
    Usingcached https://files.pythonhosted.org/packages/bf/2e/66af8d975/yagmail-
0.11.214-py2.py3-none-any.whl
Installing collected packages: yagmail
Successfully installed yagmail-0.11.214
```

1. 使用 yagmail 发送带正文的邮件

示例如下：

```
import yagmail   # 导入 yagmail 模块
# 链接邮箱服务器
yagindex = yagmail.SMTP(user="xxxxxx@qq.com",password="qq 邮箱授权码",
                        host='smtp.qq.com')
Yag_contents = [' 这是一个 yagmail 发送邮件正文的实例 ']          # 邮箱正文
yagindex.send('xxxxxx@126.com','Yagmail 主题实例 ',Yag_contents)   # 给单个接收者发送邮件
```

如果给多个接收者发送邮件，可以使用列表存储多个接收人。

```
# 给多个接收者发送邮件
yagindex.send(['aa@126.com','bb@qq.com','cc@gmail.com'],'subject',Yag_contents)
```

导入 yagmail 模块，yagmail 调用 SMTP 服务。定义发送者账号信息、客户端授权码，以及服务器的主机地址和邮件正文。最后调用 send() 方法发送给接收人，并传入邮件主题和内容。发送成功后，如图 2.44 所示。

图 2.44　邮件正文内容界面

2. 使用 yagmail 发送带附件的邮件

示例如下：

```
import yagmail    # 导入 yagmail 模块
# 链接邮箱服务器
yagindex = yagmail.SMTP(user="xxxxxx@qq.com",password="qq 邮箱授权码 ",host='smtp.qq.com')
# 邮箱正文
Yag_contents = [' 这是一个 yagmail 发送邮件正文的实例 ']
```

查看发送结果：

```
# 发送带附件的邮件
yagindex.send('xxxxxx@126.com','Yagmail 发送带附件主题 ',Yag_contents,["E://cnblogs.png"])
```

注意：如果发送多个附件，在列表内追加其他附件的所在路径即可。发送完成后，如图 2.45 所示。

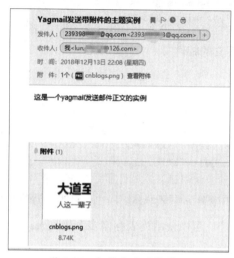

图 2.45　邮件中的附件界面

2.14　自动化测试封装实战

很多初学者在进行自动化测试过程中，前期都会把测试脚本和元素写在一起。例如，针对一个测试项目，可以多次在页面上模拟用户单击和输入等操作。当开发者修改了代码后，元素位置发生改变，需要重新修改元素定位的语法。当在编写了大量测试用例的情况下，这种维护成本就变得越来越高。本节通过自动化测试脚本的封装实战来解决这一类问题的出现。

2.14.1　自动化封装实战（一）

以登录功能为例，一般登录的整个流程是不会变化的。例如，输入用户名和密码，单击"登录"

按钮。但是对于用户名和密码，"登录"按钮对应的元素可能会出现变化。一种比较简单的方法是将登录流程的脚本封装定义为一个模板，当调用时，只需要对模板进行入参即可。先来看一个简单的线性脚本，示例如下：

```python
from selenium import webdriver
from selenium.webdriver.common.by import By                        # 导入 By 类
from selenium.webdriver.support.ui import WebDriverWait            # 导入 WebDriverWait 类
from selenium.webdriver.support import expected_conditions as EC   # 导入 EC 模块
import time
driver = webdriver.Chrome()                                        # 获取驱动
driver.get('https://www.gjfax.com/toLogin')                        # 登录测试项目
driver.refresh()
driver.implicitly_wait(30)
driver.find_element_by_id('mobilePhone').send_keys('18513600×××') # 输入用户名
element = WebDriverWait(driver,10,0.5).until(EC.presence_of_element_located((By.ID,
'password')))
element.send_keys('a123456')                                       # 显示等待输入密码
login_Btn = driver.find_element_by_id('loginBtn').click()          # 单击"登录"按钮
time.sleep(3)
driver.quit()
```

上面的测试代码是一个完整的登录流程。在测试代码中，不难发现测试数据、元素和操作流程都是放在一起的，这种方式非常不利于后期脚本的维护。对上面代码进行二次改造，示例如下：

```python
from selenium import webdriver
from selenium.webdriver.common.by import By                        # 导入 By 类
from selenium.webdriver.support.ui import WebDriverWait            # 导入 WebDriverWait 类
from selenium.webdriver.support import expected_conditions as EC   # 导入 EC 模块
import time
def Element_Locator(*new_loctor):
    '''重写 find_element 定位方法'''
    return driver.find_element(*new_loctor)
def input_username(username,*userLoctor):
    '''输入用户名'''
    return Element_Locator(*userLoctor).send_keys(username)
def input_password(password,*passwdLoctor):
    '''输入密码'''
    return Element_Locator(*passwdLoctor).send_keys(password)
def click_btn(*clickLoctor):
    '''单击"登录"按钮'''
    return Element_Locator(*clickLoctor).click()
def assert_Login_text(*assertText):
    '''获取登录成功后的验证信息'''
    return Element_Locator(*assertText).text
```

上述代码对 By 类提供的 find_element() 方法进行二次封装，封装后的定位统一写法改为 Element_Locator()，*idLoctor 表示接收的参数是不固定的，注意，所有带 * 的默认参数在传递时要放在正常参数的后面。将输入用户名、输入密码、单击"登录"按钮和获取登录成功后的验证信息分别封装成了 input_username()、input_password()、click_btn() 和 assert_Login_text() 方法，示例如下：

```python
if __name__ == '__main__':
    # 定位器
    user_loc = (By.ID,'mobilePhone')
    passwd_loc = (By.ID,'password')
    click_loc = (By.ID,'loginBtn')
    successLogin_loc = (By.CSS_SELECTOR,'a.fc-blue.mr-5')
    driver = webdriver.Chrome()                              # 获取驱动
    driver.get('https://www.gjfax.com/toLogin')              # 获取测试网址
    driver.maximize_window()                                 # 最大化窗口
    time.sleep(3)
    driver.refresh()
    input_username('18513600×××',*user_loc)                  # 输入用户名
    input_password('a123456',*passwd_loc)                    # 输入密码
    click_btn(*click_loc)                                    # 单击"登录"按钮
    time.sleep(3)
    # 断言
    if assert_Login_text(*successLogin_loc) == '安全退出':
        print('测试用例通过！')
    else:
        print('测试用例失败！')
    driver.quit()
    dr.quit()   # 关闭
```

在测试用例中调用时，分别传入对应的元素和数据，最后对登录后的结果进行判断。

2.14.2 自动化封装实战（二）

针对 2.14.1 小节的代码，可进行代码重构，即把整个登录功能结合面向对象进行再次封装，示例如下：

```python
from selenium import webdriver
from time import sleep
from selenium.webdriver.common.by import By         # 二次定位的 By 类
class GJsProject(object):
    def __init__(self):
        self.driver = webdriver.Chrome()            # 定义驱动
    def openbrowser(self,url):                      # 定义基础操作
        self.driver.get(url)
        self.driver.maximize_window()
```

```python
        sleep(2)
    def by_css(self,loc):
        ''' 重写css定位 '''
        return self.driver.find_element(By.CSS_SELECTOR,loc)
    def click_login_btn(self,loc):                          # 单击"登录"按钮
        self.by_css(loc).click()
    def input_username_Text(self,loc,text):                 # 输入用户名
        self.by_css(loc).send_keys(text)
    def input_password_Text(self,loc,text):                 # 输入密码
        self.by_css(loc).send_keys(text)
    def click_login_button(self,loc):                       # 单击"登录"按钮
        self.by_css(loc).click()
    def assert_success_text(self,loc):                      # 获取验证信息
        return self.by_css(loc).text
    def logsys_gjs_action(self,loc):                        # 退出系统
        self.by_css(loc).click()
        sleep(2)
        self.driver.quit()
```

openbrowser() 方法用于获取地址、最大化窗口基础操作。对 By 类中的 css 定位重写为 by_css() 方法。将输入用户名、输入密码、单击"登录"按钮 3 个流程封装成单独的方法。assert_success_text() 方法用于登录成功后返回文本信息，示例如下：

```python
    # 登录流程
    def login_gjs(self,url,loc1,loc2,username,loc3,password,loc4,loc5,exceptText,loc6):
        self.openbrowser(url)
        sleep(1)
        self.click_login_btn(loc1)
        sleep(1)
        self.input_username_Text(loc2,username)
        sleep(1)
        self.input_password_Text(loc3,password)
        sleep(1)
        self.click_login_button(loc4)
        sleep(1)
        if self.assert_success_text(loc5) == exceptText:    # 断言登录是否成功
            print('pass')
        else:
            print('fail')
        self.logsys_gjs_action(loc6)
```

将获取地址、最大化窗口、输入用户名、输入密码和获取验证信息等操作进一步封装为 login_gjs() 方法，这样做的好处是在调用时只需传入对应的元素和测试数据即可，示例如下：

```python
if __name__ == '__main__':
    t = GJs()
```

```
url = 'https://www.gjfax.com/'           # 打开项目地址
loc1 = "span.menubar-btn .fc-white"      # 定位器
loc2 = "# mobilePhone"
username = '18513600×××'
loc3 = "# password"
password = 'a123456'
loc4 = "# loginBtn"
loc5 = "a.fc-blue.mr-5"
exceptText = ' 安全退出 '
loc6 = "a.fc-blue.mr-5"
# 调用登录方法
t.login_gjs(url,loc1,loc2,username,loc3,password,loc4,loc5,exceptText,loc6)
```

2.15 测试框架封装和脚本的分层设计

面对一个自动化测试项目时，当编写了大量的自动化测试用例后，慢慢会发现众多问题。例如，测试用例的维护工作会变得越来越困难，其中包括页面元素经常变动，经常要修改脚本；测试数据的管理不规范；用例失败后如何发送截图通知；如何跟踪日志信息记录；计算每一次跑完用例通过数、失败数及自动发送邮件通知；构建无人值守的自动化测试任务等。

以上都是在进行自动化测试过程中需要考虑的，当然这也是自动化测试框架开发的一个思路。本节根据自动化项目实战进行自动化测试框架的开发和设计。

2.15.1 PageObject 设计模式

PageObject 简称 PO，表示页面对象。PO 设计模式可以理解为实施 Selenium 的最佳实践方式之一。它主要实现了代码的分层设计，将页面元素、元素操作和页面业务进行分离。这样做的好处是减少代码的维护量，主要表现在页面元素和元素操作的相互分离（特别是面对 UI 频繁变化的项目）；提高测试用例的可读性，使得业务测试流程更加清晰和明确。

1. PageObject 设计模式核心要素

（1）基础类：封装基础类（BasePage），基础类可以包含 WebDriver 实例的属性，如驱动的定义、对元素定位的二次封装等。

（2）页面层：每一个页面类（Page）都要继承基础类，并通过驱动（driver）来管理本页的元素，并将 Page 类中需要用到的操作都封装成一个个方法。

(3) 用例层：测试用例（TestCase）必须继承 unittest.Testcase 类，并调用相应的 Page 类来实现相应的测试步骤，最后是用例的断言。

2. 广金所登录功能 PO 分层设计实战

广金所项目登录界面如图 2.46 所示。

图 2.46　广金所项目登录界面

（1）基础类（BasePage）的实现。

```
PO\basepase\homeBase.py
PO\basepase\__init__.py
```

首先定义一个基础类，即 homeBase.py 类。homeBage.py 代码如下：

```python
from selenium.webdriver.support.wait import WebDriverWait        # 显示等待
from selenium.webdriver.support import expected_conditions as EC # 判断元素是否被定位到
class HomePage(object):                  # 页面的基础类
    def __init__(self,url,dr):           # 定义驱动和地址
        self.url = url
        self.dr = dr
    def find_element(self,*loc):         # 封装元素定位方式
        try:
            WebDriverWait(self.dr,20).until(EC.visibility_of_element_located(loc))
            return self.dr.find_element(*loc)
        except:
            print(*loc + '元素定位在页面中无法找到！')
```

在基础类的初始化方法 __init__() 中定义驱动（driver）和 URL。对 By 类的 find_element() 定位方式进行重写。如果找不到元素就抛出 except 异常。HomePage 类用于所有页面的继承。

（2）登录页面类（Page）层的实现。

```
PO\page\__init__.py
PO\page\loginpage.py
```

定义 loginpage.py 文件,将登录页面写成一个页面类,继承基础类(HomePage),示例如下:

```python
import sys
sys.path.append('../basePage')
from homePage import HomePage                          # 导入 HomePage 基础类
from selenium.webdriver.common.by import By            # 定位方式
from time import sleep
class LoginPage(HomePage):                             # 继承 HomePage 基础类
    # 定位器
    # 用户名
    username_loc = (By.ID,'mobilePhone')
    # 密码
    password_loc = (By.ID,'password')
    # "登录"按钮
    loginBtn_loc = (By.ID,'loginBtn')
    # 退出连接
    logoutBtn_loc = (By.CSS_SELECTOR,'a.fc-blue.mr-5')
    # 用户名为空
    userNull_loc = (By.CSS_SELECTOR,'# error > span')
    # 密码为空
    passWordNull_loc = (By.CSS_SELECTOR,'# error > span')
```

定义 LoginPage 页面类并继承 HomePage 基础类,分别定义登录页面上需要用到的元素。

```python
    # 打开登录页面
    def openLoginPage(self):
        self.dr.get(self.url)
        self.dr.refresh()
        self.dr.maximize_window()
        sleep(0.5)
    # 输入用户名
    def input_userName(self,userName):
        self.find_element(*self.username_loc).send_keys(userName)
    # 输入密码
    def input_passWord(self,password):
        self.find_element(*self.password_loc).send_keys(password)
    # 单击"登录"按钮
    def click_loginBtn(self):
        self.find_element(*self.loginBtn_loc).click()
    # 获取登录成功后的提示信息
    def get_assertText(self):
        return self.find_element(*self.logoutBtn_loc).text
    # 用户名为空的提示
    def get_userNullText(self):
```

```
    return self.find_element(*self.userNull_loc).text
# 密码为空的提示
def get_passwordNullText(self):
    return self.find_element(*self.passWordNull_loc).text
# 组装成登录流程
def login_gjs_pro(self,username,password):
    self.input_userName(username)
    self.input_passWord(password)
    self.click_loginBtn()
```

LoginPage 类中主要对登录页面上的元素进行封装，如将用户名、密码和"登录"按钮、验证信息等都封装成方法。login_gjs_pro() 函数将单个元素操作组成一个完整的动作，包含打开浏览器、输入用户名、密码并单击"登录"按钮等。使用时将 username、password 作为函数的入参，这样的函数具有很强的可重用性。

（3）用例层（业务层）的实现。

```
PO\testCases\__init__.py
PO\testCases\testLogin.py
```

接下来对整个登录业务进行测试用例编写。编写 testLogin.py 文件，示例如下：

```
import sys
sys.path.append('../basepage')
sys.path.append('../page')
from homePage import *      # 导入基础类
from loginpage import *     # 导入页面类
from selenium import webdriver
import unittest
from time import sleep
class TestLogin(unittest.TestCase):
    def setUp(self):
        self.url = 'https://www.gjfax.com/toLogin'
        self.dr = webdriver.Chrome()
        self.dr.implicitly_wait(30)
        # 实例化一个 loginpage 对象
        self.loginpage = LoginPage(self.url,self.dr)
    def tearDown(self):
        self.dr.quit()
    def testlogin(self):
        ''' 正确的用户名和密码 '''
        self.loginpage.openLoginPage()                                  # 打开浏览器
        self.loginpage.login_gjs_pro('18513600×××','a123456')           # 输入用户名和密码
        self.assertEqual(self.loginpage.get_assertText(),' 安全退出 ')   # 断言
    def test_user_null(self):
        ''' 测试密码为空 '''
```

```
        self.loginpage.openLoginPage()
        self.loginpage.login_gjs_pro('18513600×××','')
        self.assertEqual(self.loginpage.get_passwordNullText(),'请输入密码')   # 断言
    def test_password_null(self):
        ''' 测试用户名为空 '''
        self.loginpage.openLoginPage()
        self.loginpage.login_gjs_pro('','a123456')
        self.assertEqual(self.loginpage.get_passwordNullText(),'请输入用户名/手机号')
    def test_user_passwd_null(self):
        ''' 测试用户名/密码为空 '''
        self.loginpage.openLoginPage()
        self.loginpage.login_gjs_pro('','')
        self.assertEqual(self.loginpage.get_passwordNullText(),'请输入用户名/手机号')
if __name__ == '__main__':
    unittest.main(verbosity=2)
```

TestLogin 类继承 unitest.TestCase 类，LoginPage 类继承 HomePage 基础类，继承基础类时需要传入浏览器驱动和测试地址。LoginPage 类构造的实例化 loginpage 对象用于打开浏览器，完成测试用例的创建。

用例层不需要关注按钮、文本框的元素定位，只需要关注用哪个浏览器、登录用户名和密码数据是什么、断言数据是什么即可。

通过 PO 设计模式实现了不同层关注的问题不同，真正做到了基础类提供的方法供页面类、用例层调用，页面类关注元素操作，用例层关注业务逻辑。

2.15.2 分离测试固件

在 PO 设计模式的基础上需要对 PO 设计模式进行一次重构，首先要做的是分离测试固件。关于这部分内容，在 2.11.5 小节有详细的介绍。

在整个测试目录架构下新建一个 ownUnit.py 文件，如下所示：

```
PO\common\ownUnit.py
PO\common\__init__.py
```

编写 ownUnit.py 文件，示例如下：

```
import sys
sys.path.append('../basepage')
sys.path.append('../page')
from homePage import *
from loginpage import *
from selenium import webdriver
import unittest
from time import sleep
```

```python
class MyunitTests(unittest.TestCase):
    def setUp(self):
        self.url = 'https://www.gjfax.com/toLogin'
        self.dr = webdriver.Chrome()
        self.dr.implicitly_wait(30)
        # 实例化一个loginpage对象
        self.loginpage = LoginPage(self.url,self.dr)
    def tearDown(self):
        self.dr.quit()
```

将用例层中测试固件 setUp() 和 tearDown() 的公共部分单独定义成一个测试类 MyunitTests()，方便其他用例层调用，这样的代码设计也比较合理。

同时，修改用例层 testLogin.py 文件，示例如下：

```python
import sys,unittest
sys.path.append('../common')
sys.path.append('../page')
from ownUnit import MyunitTests      # 导入测试固件所在类 MyunitTests
class TestLogin(MyunitTests,LoginPage):
    def testlogin(self):
        ''' 正确的用户名和密码 '''
        self.loginpage.openLoginPage()
        self.loginpage.login_gjs_pro('18513600×××','a123456')
        self.assertEqual(self.loginpage.get_assertText(),' 安全退出 ')   # 断言
    def test_user_null(self):
        ''' 测试密码为空 '''
        self.loginpage.openLoginPage()
        self.loginpage.login_gjs_pro('18513600×××','')
        self.assertEqual(self.loginpage.get_passwordNullText(),' 请输入密码 ')   # 断言
    def test_password_null(self):
        ''' 测试用户名为空 '''
        self.loginpage.openLoginPage()
        self.loginpage.login_gjs_pro('','a123456')
        self.assertEqual(self.loginpage.get_passwordNullText(),' 请输入用户名 / 手机号 ')
    def test_user_passwd_null(self):
        ''' 测试用户名 / 密码为空 '''
        self.loginpage.openLoginPage()
        self.loginpage.login_gjs_pro('','')
        self.assertEqual(self.loginpage.get_passwordNullText(),' 请输入用户名 / 手机号 ')
if __name__ == '__main__':
    unittest.main(verbosity=2)
```

TestLogin 测试类继承 MyunitTests 类。通过 Myunitests 类来完成驱动的定义、浏览器获取和实例化 loginpage 对象操作。由 LoginPage 类构造的 loginpage 对象打开浏览器，完成整个登录测试业务的实现。运行结果如下：

```
test_password_null (__main__.TestLogin)
测试用户名为空 ... ok
test_user_null (__main__.TestLogin)
测试密码为空 ... ok
test_user_passwd_null (__main__.TestLogin)
测试用户名/密码为空 ... ok
testlogin (__main__.TestLogin)
正确的用户名和密码 ... ok
----------------------------------------------------------------------
Ran 4 tests in 45.961s
OK
```

2.15.3 分离测试数据

接下来对用例层的测试数据进行分离。可以使用 Excel、YAML 和 DDT 等来管理测试过程中需要使用到的测试数据，具体参见 2.12 节数据驱动测试实战。本小节使用 Excel 对测试数据进行管理。

测试架构目录中新增 data 目录，新增的 info.xlsx 表如下：

```
PO\data\info.xlsx
PO\data\__init__.py
```

针对登录界面进行自动化测试用例编写，先构造多组登录测试数据，然后将测试数据存放在 Excel 中进行管理，如图 2.47 所示。

账号	密码	预期结果
18513600×××	a123456	安全退出
18513600××××		请输入密码
	a123456	请输入用户/密码
		请输入用户/密码

图 2.47　登录测试数据展示

图 2.47 中的 4 组测试数据分别验证以下内容。

（1）用户名正确，密码正确，登录成功，"安全退出"文本验证登录成功。

（2）用户名错误，密码为空，登录失败，"请输入密码"文本验证登录失败。

（3）用户名为空，密码正确，登录失败，"请输入用户名/密码"文本验证登录失败。

（4）用户名为空，密码为空，登录失败，"请输入用户名/密码"文本验证登录失败。

在测试目录架构中新增 helper.py 文件，即 PO\common\helper.py，示例如下：

```
import xlrd
import logging,os
class Helper(object):
```

```python
def readExcles(self,rowx):
    '''
    rowx 是行数
    :param rowx:
    :return: 返回的每一个行的行数
    '''
    book = xlrd.open_workbook(r'...\data\info.xlsx','r')
    table = book.sheet_by_index(0)
    return table.row_values(rowx)
def readusername(self,rowx):
    '''
    rowx 返回的是第几行的用户名
    :param rowx:
    :return:
    '''
    return str(self.readExcles(rowx)[0])
def readpassword(self,rowx):
    '''
    rowx 返回的是第几行的密码
    :param rowx:
    :return:
    '''
    return self.readExcles(rowx)[1]
def exceptText(self,rowx):
    '''
    rowx 返回的是第几行的预期结果
    :param rowx:
    :return:
    '''
    return self.readExcles(rowx)[2]
def dirname(self,filename,filepath='data'):
    '''
    :param filename: 文件名
    :param filepath: 文件路径
    :return:
    '''
    return os.path.join(os.path.dirname(os.path.dirname(__file__)),filepath,filename)
```

readExcles() 方法用来返回 Excel 中的第几行数据，readusername() 方法表示读取第几行的第一列数据，readpassword() 方法表示读取第几行的第二列数据，exceptText() 方法表示读取第几行的第三列数据。这样设计的好处是只需要关注 Excel 中第几行，不需要关注第几列，就可以获取到指定的登录账号、密码和预期结果。driname() 方法用来读取 info.xlxs 表中的数据。

修改 testLogin.py 文件，示例如下：

```python
import sys,unittest
sys.path.append('../common')
sys.path.append('../page')
from ownUnit import MyunitTests              # 导入测试关键所在类
from helper import Helper                    # 新增 Helper 类
from time import sleep
class TestLogin(MyunitTests,Helper):
    def testlogin(self):
        ''' 正确的用户名和密码 '''
        self.loginpage.openLoginPage()   # 打开项目首页
        # 测试账号和密码
        self.loginpage.login_gjs_pro(self.readusername(1),self.readpassword(1))
        self.assertEqual(self.loginpage.get_assertText(),self.exceptText(1))   # 断言
    def test_user_null(self):
        ''' 测试密码为空 '''
        self.loginpage.openLoginPage()
        self.loginpage.login_gjs_pro(self.readusername(2),self.readpassword(2))
        # 断言
        self.assertEqual(self.loginpage.get_passwordNullText(),self.exceptText(2))
    def test_password_null(self):
        ''' 测试用户名为空 '''
        self.loginpage.openLoginPage()
        self.loginpage.login_gjs_pro(self.readusername(3),self.readpassword(3))
        self.assertEqual(self.loginpage.get_userNullText(),self.exceptText(3))
    def test_user_passwd_null(self):
        ''' 测试用户名/密码为空 '''
        self.loginpage.openLoginPage()
        self.loginpage.login_gjs_pro(self.readusername(4),self.readpassword(4))
        self.assertEqual(self.loginpage.get_passwordNullText(),self.exceptText(4))
if __name__ == '__main__':
    unittest.main(verbosity=2)
```

TestLogin 类继承 Helper 类。readusername() 方法和 readpassword() 方法用于读取 Excel 中的登录用户名和密码，exceptText() 方法用于读取 Excel 中的预期结果，调用时只需传入要获取第几行即可。

2.15.4 用例失败截图

有时自动化测试用例在执行过程中可能会出现失败的情况，这种情况可以通过引入截图来定位分析问题，下面继续优化 PO 设计模式。

测试目录架构新增 getImage.py 文件，即 PO\common\getImage.py。编写 getImage.py 文件，示例如下：

```python
import os,time
def SaveImage(driver,errorImage):
    ''' 用例失败截图功能 '''
    Rawpath = os.path.join(os.path.dirname(os.path.dirname(__file__)),'Image')
    NewPicture = Rawpath + '\\' + time.strftime('%Y_%y_%d_%H_%M_%S') + '_' + errorImage
    driver.get_screenshot_as_file(NewPicture)
```

Rawpath 定义了 Image 目录的绝对路径,目的是将生成的截图写入了该目录下。time.strftime() 方法返回的是一个字符串时间,用来区别不同的截图名称;get_screenshot_as_file() 方法用于生成用例成功或失败截图。修改 testLogin.py 文件,示例如下:

```python
import sys,unittest
sys.path.append('../common')
sys.path.append('../page')
from ownUnit import MyunitTests          # 导入测试关键所在类
from helper import Helper                # 新增 Helper 类
from time import sleep
from getImage import SaveImage           # 导入截图功能
class TestLogin(MyunitTests,Helper):
    def testlogin(self):
        ''' 正确的用户名和密码 '''
        self.loginpage.openLoginPage()
        self.loginpage.login_gjs_pro(self.readusername(1),self.readpassword(1))
        self.assertEqual(self.loginpage.get_assertText(),self.exceptText(1))  # 断言
        SaveImage(self.dr,'login_success.png')
    def test_user_null(self):
        ''' 测试密码为空 '''
        self.loginpage.openLoginPage()
        self.loginpage.login_gjs_pro(self.readusername(2),self.readpassword(2))
        # 断言
        self.assertEqual(self.loginpage.get_passwordNullText(),self.exceptText(2))
        SaveImage(self.dr,'loginpasswdNull.png')
    def test_password_null(self):
        ''' 测试用户名为空 '''
        self.loginpage.openLoginPage()
        self.loginpage.login_gjs_pro(self.readusername(3),self.readpassword(3))
        self.assertEqual(self.loginpage.get_userNullText(),self.exceptText(3))
        SaveImage(self.dr,'loginuserNull.png')
    def test_user_passwd_null(self):
        ''' 测试用户名/密码为空 '''
        self.loginpage.openLoginPage()
        self.loginpage.login_gjs_pro(self.readusername(4),self.readpassword(4))
        self.assertEqual(self.loginpage.get_passwordNullText(),self.exceptText(4))
        SaveImage(self.dr,'loginuserAndpasswd.png')
if __name__ == '__main__':
    unittest.main(verbosity=2)
```

导入 SaveImage 截图功能，分别在每一条测试用例执行完成后自动截图。调用 SaveImage() 方法时需传入 dr 驱动和图片格式，生成的用例截图如图 2.48 所示。

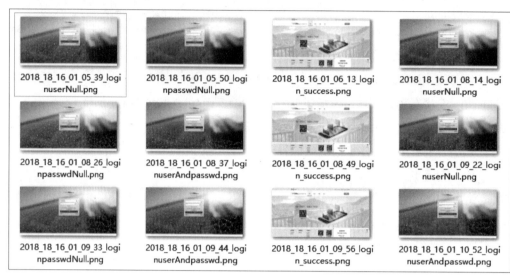

图 2.48　生成的用例截图

2.15.5　日志跟踪收集

logging 模块是 Python 的一个标准库，可以用来标注日志记录，更方便地进行日志记录，同时还可以做更方便的级别区分，以及一些额外日志信息的记录，如时间和运行模块信息等。

1. logging 模块的基本使用示例

示例如下：

```
import logging
logging.basicConfig(level=logging.INFO,format='%(asctime)s-%(name)s-%(levelname)s-
                %(message)s')
logger = logging.getLogger(__name__)
logger.info("开始：输出 Logging 日志")
logger.debug("这是显示 Debug 级别的日志信息")
logger.warning("这是显示 Warning 级别的日志信息")
logger.info("结束：输出 Logging 日志")
```

输出结果如下：

```
2018-12-16 11:37:51,251 - __main__ - INFO - 开始：输出 Logging 日志
2018-12-16 11:37:51,251 - __main__ - WARNING - 这是显示 Warning 级别的日志信息
2018-12-16 11:37:51,251 - __main__ - INFO - 结束：输出 Logging 日志
```

logging.basicConfig() 函数的参数分别如下：level 参数表示日志等级，一般常见的有 info、

debug、warning 和 error 等；format 参数输出指定的日期格式和内容，可以使日志信息变得更加详细。

level 参数可以设置不同的日志等级，用于控制日志的输出，说明如下。

（1）FATAL：致命错误。

（2）INFO：处理请求或状态变化等日常事务。

（3）DEBUG：调试过程中使用 DEBUG 等级，如算法中每个循环的中间状态。

（4）CRITICAL：特别糟糕的事情，如内存耗尽、磁盘空间为空，一般很少使用。

（5）ERROR：发生错误，如 IO 操作失败或连接问题。

（6）WARNING：发生很重要的事件，但是并不是错误，如用户登录密码错误。

2. 将 logging 日志写入文件中

设置 logging，创建一个 FileHandler() 对象，并对输出消息的格式进行设置，将其添加到 logger，然后将日志写入指定的文件中，示例如下：

```python
import logging
# 定义文件
logFile = logging.FileHandler('logTest.txt','a',encoding='utf-8')
# 设置 log 日志格式
fmt = logging.Formatter(fmt='%(asctime)s-%(name)s-%(levelname)s-
                            %(module)s:%(message)s')
logFile.setFormatter(fmt)
# 定义日志级别
LoggerMany = logging.Logger('logTest',level=logging.DEBUG)   # 设置日志级别
LoggerMany.addHandler(logFile)
# 写入内容到 logging 日志
LoggerMany.critical('info')
# 输出日志信息
LoggerMany.info("info 级别的信息日志 ")
LoggerMany.debug("debug 级别的信息日志 ")
LoggerMany.warning("warning 级别的信息日志 ")
LoggerMany.info("info 级别的信息日志 ")
```

FileHandler() 对象中，a 表示对日志文件信息进行追加；encoding="utf-8" 设置中文编码，防止乱码。创建 logTest.txt 日志文件，如图 2.49 所示。

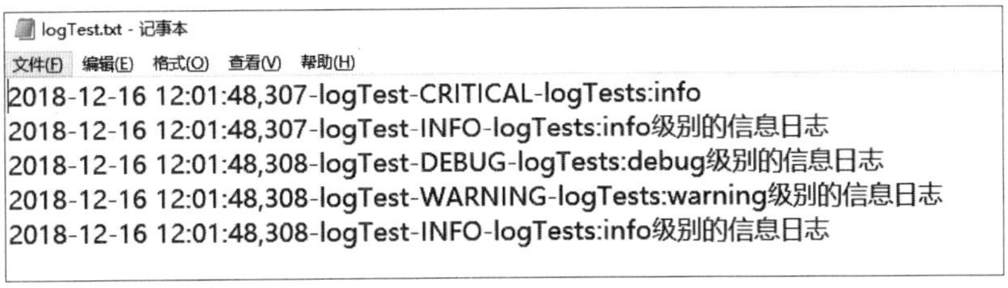

```
logTest.txt - 记事本
文件(F)  编辑(E)  格式(O)  查看(V)  帮助(H)
2018-12-16 12:01:48,307-logTest-CRITICAL-logTests:info
2018-12-16 12:01:48,307-logTest-INFO-logTests:info级别的信息日志
2018-12-16 12:01:48,308-logTest-DEBUG-logTests:debug级别的信息日志
2018-12-16 12:01:48,308-logTest-WARNING-logTests:warning级别的信息日志
2018-12-16 12:01:48,308-logTest-INFO-logTests:info级别的信息日志
```

图 2.49　logTest.txt 日志文件

3. logging 日志在自动化测试中的应用

在测试目录架构 PO\common\helper.py 中，打开 helper.py 文件并修改代码。新增 dirname() 方法，用于生成日志文件，示例如下：

```python
def dirname(self,filename,filepath='data'):
    '''
    :param filename: 文件名
    :param filepath: 文件路径
    :return:
    '''
    return os.path.join(os.path.dirname(os.path.dirname(__file__)),filepath,filename)
def log(self,log_content):
    ''' 定义 log 日志级别 '''
    # 定义日志文件
    logFile = logging.FileHandler(self.dirname('log.txt'),'a',encoding='utf-8')
    # 设置 log 格式
    fmt = logging.Formatter(fmt='%(asctime)s-%(name)s-%(levelname)s-'
                                '%(module)s:%(message)s')
    logFile.setFormatter(fmt)
    logger1 = logging.Logger('logTest',level=logging.DEBUG)    # 定义日志
    logger1.addHandler(logFile)
    logger1.info(log_content)
    logFile.close()
```

log() 函数的作用是设置日志级别，并将日志信息写入 PO\data\ 目录下。在调用 dirname() 函数时需要传入日志文件名。接下来，修改 testLogin.py 文件，示例如下：

```python
import sys,unittest
sys.path.append('../common')
sys.path.append('../page')
from ownUnit import MyunitTests          # 导入测试关键所在类
from helper import Helper                # 新增 Helper 类
from time import sleep
from getImage import SaveImage           # 导入截图功能
import logging                           # 导入日志模块
class TestLogin(MyunitTests,Helper):
    def testlogin(self):
        ''' 正确的用户名和密码 '''
        self.loginpage.openLoginPage()
        self.log('PO-gjs：打开浏览器进入到项目首页 ')
        self.loginpage.login_gjs_pro(self.readusername(1),self.readpassword(1))
        self.log('PO-gjs：输入正确的用户名和密码 ')
        self.assertEqual(self.loginpage.get_assertText(),self.exceptText(1))  # 断言
        self.log('PO-gjs：登录成功获取信息进行断言 ')
```

```python
            SaveImage(self.dr,'login_success.png')
            self.log('PO-gjs：登录成功后获取截图信息 ')
            self.log('PO-gjs：第四条用例执行结束.....')
    def test_user_null(self):
        '''测试密码为空'''
        self.loginpage.openLoginPage()
        self.log('PO-gjs：打开浏览器进入到项目首页 ')
        self.loginpage.login_gjs_pro(self.readusername(2),self.readpassword(2))
        self.log('PO-gjs：输入正确用户名和密码为空 ')
        # 断言
        self.assertEqual(self.loginpage.get_passwordNullText(),self.exceptText(2))
        self.log('PO-gjs：登录失败获取信息进行断言 ')
        SaveImage(self.dr,'loginpasswdNull.png')
        self.log('PO-gjs：登录失败后获取截图信息 ')
        self.log('PO-gjs：第一条用例执行结束.....')
    def test_username_null(self):
        '''测试用户名为空'''
        self.loginpage.openLoginPage()
        self.log('PO-gjs：打开浏览器进入到项目首页 ')
        self.loginpage.login_gjs_pro(self.readusername(3),self.readpassword(3))
        self.log('PO-gjs：输入用户名为空和正确密码 ')
        self.assertEqual(self.loginpage.get_userNullText(),self.exceptText(3))
        self.log('PO-gjs：登录失败获取信息进行断言 ')
        SaveImage(self.dr,'loginuserNull.png')
        self.log('PO-gjs：登录失败后获取截图信息 ')
        self.log('PO-gjs：第三条用例执行结束.....')
    def test_user_passwd_null(self):
        '''测试用户名/密码为空'''
        self.loginpage.openLoginPage()
        self.log('PO-gjs：打开浏览器进入到项目首页 ')
        self.loginpage.login_gjs_pro(self.readusername(4),self.readpassword(4))
        self.log('PO-gjs：输入用户名为空和正确为空 ')
        self.assertEqual(self.loginpage.get_passwordNullText(),self.exceptText(4))
        self.log('PO-gjs：登录失败获取信息进行断言 ')
        SaveImage(self.dr,'loginuserAndpasswd.png')
        self.log('PO-gjs：登录失败后获取截图信息 ')
        self.log('PO-gjs：第二条用例执行结束.....')
if __name__ == '__main__':
    unittest.main(verbosity=2)
```

运行结果如图 2.50 所示。

```
logData.txt - 记事本
文件(F) 编辑(E) 格式(O) 查看(V) 帮助(H)
2018-12-16 12:55:23,767-logTest-INFO-helper:PO-gjs：打开浏览器进入到项目首页
2018-12-16 12:55:24,234-logTest-INFO-helper:PO-gjs：输入正确用户名和密码为空
2018-12-16 12:55:24,308-logTest-INFO-helper:PO-gjs：登录失败获取信息进行断言
2018-12-16 12:55:24,769-logTest-INFO-helper:PO-gjs：登录失败后获取截图信息
2018-12-16 12:55:24,773-logTest-INFO-helper:PO-gjs：第1条用例执行结束.....
2018-12-16 12:55:35,093-logTest-INFO-helper:PO-gjs：打开浏览器进入到项目首页
2018-12-16 12:55:35,428-logTest-INFO-helper:PO-gjs：输入用户名为空和正确为空
2018-12-16 12:55:35,494-logTest-INFO-helper:PO-gjs：登录失败获取信息进行断言
2018-12-16 12:55:36,026-logTest-INFO-helper:PO-gjs：登录失败后获取截图信息
2018-12-16 12:55:36,028-logTest-INFO-helper:PO-gjs：第2条用例执行结束.....
2018-12-16 12:55:46,244-logTest-INFO-helper:PO-gjs：打开浏览器进入到项目首页
2018-12-16 12:55:46,611-logTest-INFO-helper:PO-gjs：输入用户名为空和正确密码
2018-12-16 12:55:46,685-logTest-INFO-helper:PO-gjs：登录失败获取信息进行断言
2018-12-16 12:55:47,205-logTest-INFO-helper:PO-gjs：登录失败后获取截图信息
2018-12-16 12:55:47,206-logTest-INFO-helper:PO-gjs：第3条用例执行结束.....
2018-12-16 12:55:57,541-logTest-INFO-helper:PO-gjs：打开浏览器进入到项目首页
2018-12-16 12:55:57,941-logTest-INFO-helper:PO-gjs：输入正确的用户名和密码
2018-12-16 12:55:59,038-logTest-INFO-helper:PO-gjs：登录成功获取信息进行断言
2018-12-16 12:55:59,855-logTest-INFO-helper:PO-gjs：登录成功后获取截图信息
2018-12-16 12:55:59,857-logTest-INFO-helper:PO-gjs：第4条用例执行结束.....
```

图 2.50　logData.txt 日志文件

2.15.6　生成 HTML 格式的测试报告

引入 HTMLTestRunner 模块来统计测试用例的通过率。首先配置 HTMTestRunner 模块。将 HTMLTestRunner.py 文件放在 PO\common\ 目录下。在 PO 目录下新增 allTests.py 文件，目录结构为 PO\allTests.py，示例如下：

```
import sys,os,time,unittest
sys.path.append('./common')
sys.path.append('./basepage')
sys.path.append('./page')
from HTMLTestRunner import HTMLTestRunner
def getAllCases():
    ''' 获取 testCase 下的所有测试模块 '''
    Testsuite = unittest.defaultTestLoader.discover(
        start_dir = os.path.join(os.path.dirname(__file__),'testCases'),
        pattern = 'test*.py')
    return Testsuite
def RunMain():
    ''' 生成测试报告写入指定 Reports 目录 '''

    fp = open(os.path.join(os.path.dirname(__file__),'report',
            time.strftime("%Y_%m_%d_%H_%M_%S")+'report.html'),'wb')
    HTMLTestRunner(stream=fp,title='Python+Selenium 自动化测试实战 ',
                description=' 基于 python 语言 PO 自动化测试 ').run(getAllCases())
if __name__ == '__main__':
    RunMain()
```

运行结果如图 2.51 所示。

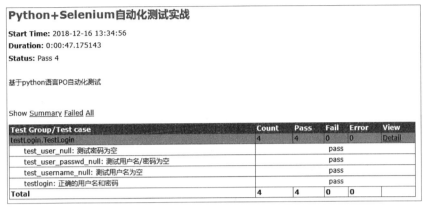

图 2.51 测试报告截图

通过测试报告可以很清楚地看见测试开始时间、结束时间、环境及失败原因等。此外，比较重要的参考信息还有测试用例通过率的用例总数、通过数和失败数。

2.15.7 发送带附件的测试报告

有时需要将生成的测试报告通过邮件发送给相关人，关于这部分的学习可以参考 2.13.2 小节带附件的邮件实践。继续优化 PO\allTests.py 文件，示例如下：

```python
import time,os,unittest,sys
sys.path.append('./common')
sys.path.append('./basepage')
sys.path.append('./page')
from HTMLTestRunner import HTMLTestRunner      # 导入生成邮件测试模版
from homePage import *                          # 导入基础类
from loginpage import *                         # 导入页面类
import smtplib                                  # 邮箱服务器
from email.mime.text import MIMEText            # 邮件模板类
from email.mime.multipart import MIMEMultipart  # 邮件附件类
from email.header import Header                 # 邮件头部模板
# 发送带邮件的函数动作
def send_mail(file_new):
    f = open(file_new,'rb')
    mail_body = f.read()
    f.close()
    # 基本信息
    smtpserver = 'smtp.126.com'
    pwd = "xxxxxxxxx"      # 126 邮箱授权码
    # 定义邮件主题
    msg = MIMEMultipart()
```

```python
msg['subject'] = Header(u'Page Object 自动化测试报告 ','utf-8')
msg['from'] = "xxxxxx@126.com"         # 必须加，不加报错。发送者邮箱账号
msg['to'] = "xxxxxx@126.com"           # 必须加，不加报错。接收者邮箱账号
# 不加 msg['from']、msg['to'] 报错原因，是因为发件人和收件人参数没有进行定义
# HTML 邮件正文
body = MIMEText(mail_body,"html","utf-8")
msg.attach(body)
att = MIMEText(mail_body,"base64","utf-8")
att["Content-Type"] = "application/octet-stream"
att["Content-Disposition"] = 'attachment;filename="test_report.html"'
msg.attach(att)
# 链接邮箱服务器发送邮件
smtp = smtplib.SMTP()
smtp.connect(smtpserver)
smtp.login(msg['from'],pwd)
smtp.sendmail(msg['from'],msg['to'],msg.as_string())
print(" 邮件发送成功 ")
```

send_mail() 函数定义了邮件主题、正文定义和链接邮箱服务器等操作。在调用该方法时，只需要传入最新的测试报告名即可，示例如下：

```python
# 查找最新邮件
def new_file(test_dir):
    result_dir = test_dir
    lists = os.listdir(result_dir)   # print(lists)  # 列出测试报告目录下的所有文件
    lists.sort()                                    # 从小到大排序 文件
    file = [x for x in lists if x.endswith('.html')] # for 循环遍历以 .html 格式的测试报告
    file_path = os.path.join(result_dir,file[-1])   # 找到测试报告目录下最新的测试报告
    return file_path                                # 返回最新的测试报告
if __name__ == '__main__':
    base_dir = os.path.dirname(os.path.realpath(__file__))      # 获取文件所在路径
    test_dir = os.path.join(base_dir,'testCases')               # 测试用例所在目录
    test_report = os.path.join(base_dir,'report')               # 测试报告所在目录
    testlist = unittest.defaultTestLoader.discover(test_dir,pattern='test*.py')
    now = time.strftime("%Y-%m-%d %H_%M_%S")
    filename = test_report + "\\" + now + 'result.html'
    fp = open(filename,'wb')
    runner = HTMLTestRunner(stream=fp,
                            title='PageObject 自动化测试报告 ',
                            description=u' 系统环境 :Win10 浏览器 :Chrome 用例执行情况 :')
    runner.run(testlist)
    fp.close()
    new_report = new_file(test_report)              # 获取最新报告文件
    send_mail(new_report)                           # 发送最新的测试报告
```

new_file() 方法用于获取最新的测试报告，在调用该方法时需要将测试报告 test_report 作为参

数传入。首先列出测试报告所在目录下的所有文件。os.listdir() 方法返回的对象是 list 对象；使用 sort() 方法对所有文件由小到大排序；列表解析式用于判断所有文件中以 .html 结尾的文件；[-1] 表示获取到最后一个 .html 文件，即最新生成的测试报告。最后将最新的测试报告 new_report 作为参数传递给 send_mail() 方法，从而实现带附件的测试报告的发送。发送后的结果如图 2.52 所示。

图 2.52　带附件的测试报告界面

2.15.8　项目持续集成

Jenkins 是一个独立的开源自动化服务器，可用于自动执行与构建、测试、交付或部署软件相关的各种任务。例如，检出代码、编译构建、运行测试、结果记录和测试统计等都是自动完成的，无须人工干预，有利于减少重复过程，以节省时间、费用和工作量。本小节通过自动化测试项目与 Jenkins 进行持续集成，构建无人值守的自动化测试任务。

因为 Jenkins 是 Java 语言开发的，所以使用 Jenkins 前必须先安装 Java 环境。JDK 可以安装 1.8 以上版本，然后配置 Java 环境变量即可。

1. 安装 Jenkins

Jenkins 的官网地址为 https://jenkins.io，本小节下载的是 Jenkins 2.95 版本 mis 包，如图 2.53 所示。

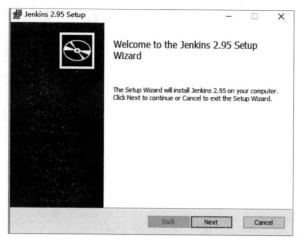

图 2.53　Jenkins 安装首页界面

安装完成后直接弹出浏览器界面，默认访问 localhost:8080 地址，如图 2.54 所示。

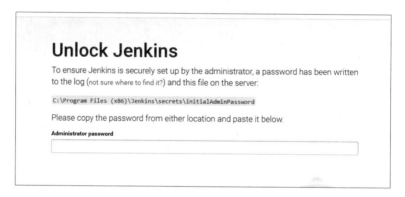

图 2.54　Jenkins 配置界面

输入管理员密码，复制 C:\Program Files (x86)\Jenkins\secrets\initialAdminPassword 到本地 C 盘路径，如图 2.55 和图 2.56 所示。

图 2.55　C 盘路径界面

图 2.56　管理员密码界面

将密码复制到 Administrator password 位置处，单击"Continue"按钮。一般建议安装系统推荐的插件，如图 2.57 所示，单击"Install suggested plugins"按钮进行安装。

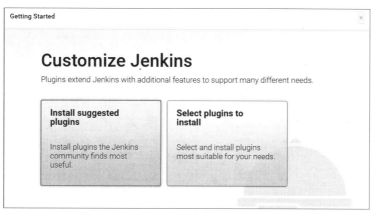

图 2.57 推荐插件安装界面

设置 Jenkins 登录账号和密码，如图 2.58 所示。

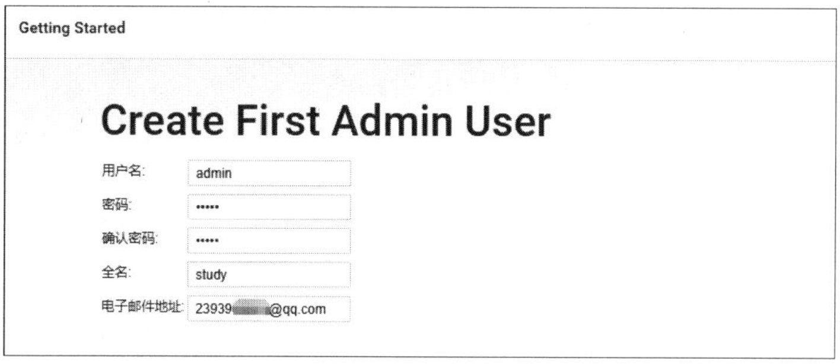

图 2.58 设置 Jenkins 登录账号和密码界面

Jenkins 登录界面如图 2.59 所示。

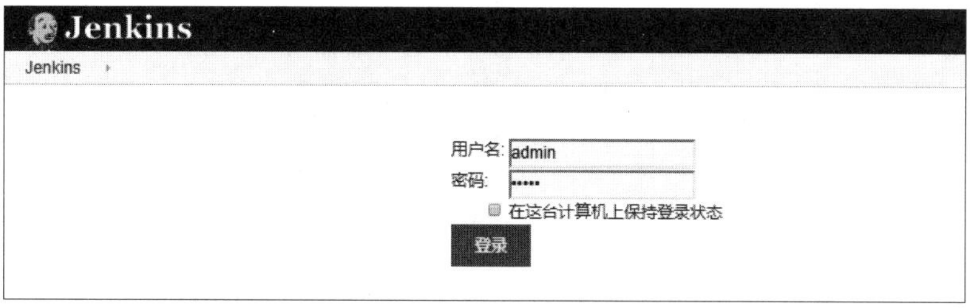

图 2.59 Jenkins 登录界面

至此，Jenkins 的配置就完成了。

2. 构建一个自由风格的任务

在 Jenkins 主界面，选择"新建任务"选项，输入一个任务名称，选择"构建一个自由风格的软件项目"选项，单击"确定"按钮。如图 2.60 所示。

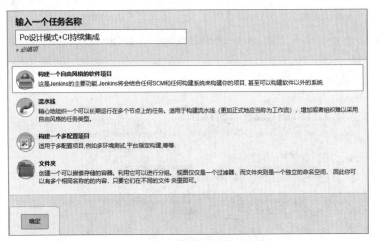

图 2.60　构建任务界面

切换到"构建"选项,在"Execute Windows batch command"界面中的"命令"文本框中输入命令行构建命令,最后单击"保存"按钮,如图 2.61 所示。

图 2.61　命令行构建界面

返回 Jenkins 主界面,选择"立即构建"选项,如图 2.62 和图 2.63 所示。

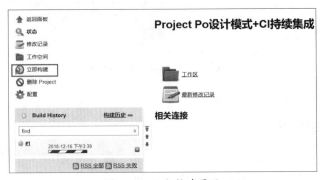

图 2.62　立即构建界面

图 2.63　构建完成界面

3. 使用 HTML Publisher plugin 实现 HTML 文档报告展示

在 Jenkins 主界面选择"系统管理"→"管理插件"→"可选插件"选项，搜索 HTML 关键字，找到可选插件中的 HTML Publisher 插件，单击直接安装，这里使用的版本是 1.17，如图 2.64 所示。安装完成后，需要重启 Jenkins 才可以生效。

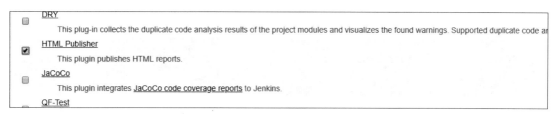

图 2.64　HTML Publisher 插件界面

注意：从版本 1.625.3 和 1.641 开始，Jenkins 限制了在提供静态文件时可以显示的内容类型，这可能会影响使用此插件存档的 HTML 文件的显示方式。

在"配置"界面，选择"构建后操作"选项卡，填入生成的测试报告所在路径，如图 2.65 所示。

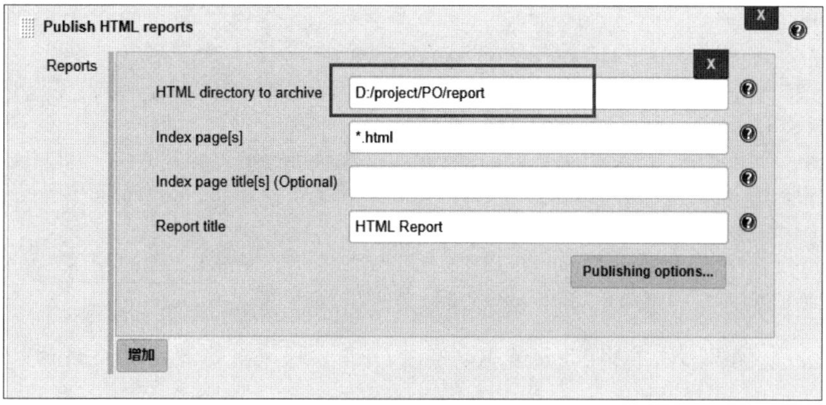

图 2.65　构建后操作界面

Reports 界面各个参数的含义如下。

（1）HTML directory to archive：指定 HTML 测试报告生成的所在目录（注意，是向右"/"的路径）。

（2）Index page[s]：生成的测试报告扩展名以 .html 结尾。

（3）Report title：HTML 测试报告的主题。

配置"构建后操作"后，保存后重新选择"立即构建"选项，如图 2.66 所示。

图 2.66　HTML 列表界面

打开 HTML 测试报告，发现如下报错：

```
Traceback (most recent call last):
  File "./common\ownUnit.py",line 16,in setUp
    self.dr = webdriver.Chrome()
  File "C:\Python36\lib\site-packages\selenium\webdriver\chrome\webdriver.py",line 81,in __init__
    desired_capabilities=desired_capabilities)
  File "C:\Python36\lib\site-packages\selenium\webdriver\remote\webdriver.py",line 157,in __init__
    self.start_session(capabilities,browser_profile)
  File "C:\Python36\lib\site-packages\selenium\webdriver\remote\webdriver.py",line 252,in start_session
    response = self.execute(Command.NEW_SESSION,parameters)
  File "C:\Python36\lib\site-packages\selenium\webdriver\remote\webdriver.py",line 321,in execute
    self.error_handler.check_response(response)
  File "C:\Python36\lib\site-packages\selenium\webdriver\remote\errorhandler.py",line 242,in check_response
    raise exception_class(message,screen,stacktrace)
selenium.common.exceptions.WebDriverException: Message: unknown error: cannot find Chrome binary
  (Driver info: chromedriver=2.38.552522 (437e6fbedfa8762dec75e2c5b3ddb86763dc9dcb),platform=Windows NT 10.0.17134 x86_64)
```

解决办法：将测试脚本中的 Chrome 驱动换成 FireFox 驱动，修改 ownUnit.py 文件，示例如下：

```
class MyunitTests(unittest.TestCase):
    def setUp(self):
```

```
        self.url = 'https://www.gjfax.com/toLogin'
        self.dr = webdriver.Firefox()    # 将 Chrome 驱动换成 Firefox 驱动
        self.dr.implicitly_wait(30)
        # 实例化一个 loginpage 对象
        self.loginpage = LoginPage(self.url,self.dr)
    def tearDown(self):
        self.dr.quit()
```

再次选择"立即构建"选项，查看测试报告，如图 2.67 所示。

图 2.67　HTML 测试报告界面

2.15.9　Jenkins+Allure 配置自动化测试报告

Allure 是一个测试报告框架，支持多语言、多平台，可以通过 Junit、TestNG 和 Pytest 框架等产生测试结果并生成酷炫好看的 Report；同时也支持自定义字段，将想展示的结果展示在报告中。本小节借助 Allure 框架来生成 HTML 测试报告。

1．Allure 环境搭建配置

首先，安装 Allure 插件。在 Jenkins 主界面选择"系统管理"→"管理插件"→"可选插件"选项，搜索 allure，从搜索结果中选中 Allure 前的复选框直接安装，如图 2.68 所示。

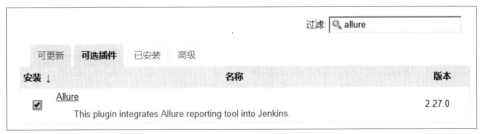

图 2.68　Allure 安装界面

其次，配置 Global Tool Configuration。在 Jenkins 主界面选择"系统管理"→"全局工具配置"选项，从中找到 JDK，单击"JDK 安装"按钮，配置 JDK 环境，如图 2.69 所示。

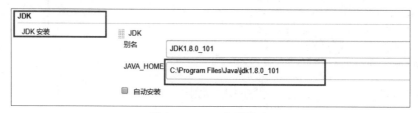

图 2.69　JDK 配置界面

最后，单击"Allure Commandline 安装"按钮，安装 Allure，建议选择 2.4.1 版本，如图 2.70 所示。选择好后，单击"保存"按钮即可。

图 2.70　Allure 配置界面

2．系统设置配置

在 Jenkins 主界面选择"系统管理"→"系统设置"选项，找到 Allure Report 后，进行配置，如图 2.71 所示。

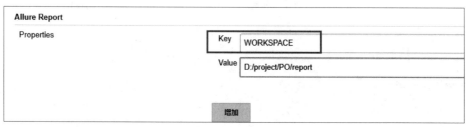

图 2.71　Allure Report 配置界面

需要说明的是，Key 的值必须是 WORKSPACE，Value 对应的值是存放 Allure 报告的路径，这个可以自定义。配置完成后，需要单击"保存"按钮。

3．配置 Job，构建后操作新增 Allure Report

在 Jenkins 主界面选择"配置"选项，切换到"构建后操作"界面，选择"Allure Report"选项，在"Path"文本框中输入"report"，如图 2.72 所示。

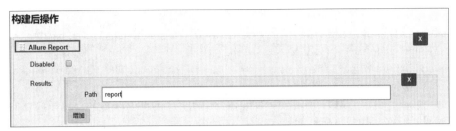

图 2.72　配置 Job 中的 Allure Report 界面

4．安装 Pytest 单元测试框架

Allure 需要结合 Pytest 才可以生成测试报告，所以需要安装 Pytest 框架，可以直接使用 pip 命令在线安装。打开 cmd 命令提示符界面，输入 "pip install pytest" 进行在线安装，示例如下：

```
C:\Users\23939>pip install pytest
Collecting pytest
Using cached https://files.pythonhosted.org/packages/19/80/1ac71d332302a89e86374560
62186bf397abc5a5b663c1919b73f4d68b1b/pytest-4.0.2-py2.py3-none-any.whl
Requirement already satisfied: setuptools in c:\python36\lib\site-packages (from
pytest) (28.8.0)
Requirement already satisfied: colorama;sys_platform == "win32" in c:\python36\lib\
site-packages (from pytest) (0.3.9)
Requirement already satisfied: atomicwrites>=1.0 in c:\python36\lib\site-packages
(from pytest) (1.1.5)
Requirement already satisfied: more-itertools>=4.0.0 in c:\python36\lib\site-
packages (from pytest) (4.2.0)
Requirement already satisfied: py>=1.5.0 in c:\python36\lib\site-packages (from
pytest) (1.5.4)
Requirement already satisfied: pluggy>=0.7 in c:\python36\lib\site-packages (from
pytest) (0.8.0)
Requirement already satisfied: attrs>=17.4.0 in c:\python36\lib\site-packages (from
pytest) (18.1.0)
Requirement already satisfied: six>=1.10.0 in c:\python36\lib\site-packages (from
pytest) (1.11.0)
Installing collected packages: pytest
Successfully installed pytest-4.0.2
```

继续使用 pip 命令安装 pytest-allure-adaptor，示例如下：

```
C:\Users\23939>pip install pytest-allure-adaptor
Collecting pytest-allure-adaptor
  Using cached https://files.pythonhosted.org/packages/2e/94/862ca2f86f3644fd6687e25
4518ff57fe729676172ef37594913e88a2e3c/pytest_allure_adaptor-1.7.10-py3-none-any.whl
Collecting enum34 (from pytest-allure-adaptor)
  Using cached https://files.pythonhosted.org/packages/af/42/cb9355df32c69b553e72a2e
28daee25d1611d2c0d9c272aa1d34204205b2/enum34-1.1.6-py3-none-any.whl
Requirement already satisfied: namedlist in c:\python36\lib\site-packages (from
```

```
pytest-allure-adaptor) (1.7)
Requirement already satisfied: six>=1.9.0 in c:\python36\lib\site-packages (from
pytest-allure-adaptor) (1.11.0)
Requirement already satisfied: pytest>=2.7.3 in c:\python36\lib\site-packages (from
pytest-allure-adaptor) (4.0.2)
Requirement already satisfied: lxml>=3.2.0 in c:\python36\lib\site-packages (from
pytest-allure-adaptor) (4.2.3)
Requirement already satisfied: setuptools in c:\python36\lib\site-packages (from
pytest>=2.7.3->pytest-allure-adaptor) (28.8.0)
Requirement already satisfied: attrs>=17.4.0 in c:\python36\lib\site-packages (from
pytest>=2.7.3->pytest-allure-adaptor) (18.1.0)
Requirement already satisfied: more-itertools>=4.0.0 in c:\python36\lib\site-
packages (from pytest>=2.7.3->pytest-allure-adaptor) (4.2.0)
Requirement already satisfied: pluggy>=0.7 in c:\python36\lib\site-packages (from
pytest>=2.7.3->pytest-allure-adaptor) (0.8.0)
Requirement already satisfied: colorama;sys_platform == "win32" in c:\python36\lib\
site-packages (from pytest>=2.7.3->pytest-allure-adaptor) (0.3.9)
Requirement already satisfied: atomicwrites>=1.0 in c:\python36\lib\site-packages
(from pytest>=2.7.3->pytest-allure-adaptor) (1.1.5)
Requirement already satisfied: py>=1.5.0 in c:\python36\lib\site-packages (from
pytest>=2.7.3->pytest-allure-adaptor) (1.5.4)
Installing collected packages: enum34,pytest-allure-adaptor
Successfully installed enum34-1.1.6 pytest-allure-adaptor-1.7.10
```

5. 配置Job信息

修改"构建环境"选项卡中的信息，修改完成后单击"保存"按钮，如图2.73所示。

图2.73 配置Job界面

"命令"文本框中的信息含义如下。

（1）D：表示切换到 D 盘。

（2）cd D:/project/PO/testCases：切换到测试用例所在目录位置。

（3）python -m pytest　--alluredir ${WORKSPACE}/report：生成 Allure 报告并写入 report 目录下。

6．立即构建，查看 Allure 测试报告

控制台输出结果，如图 2.74～图 2.77 所示。

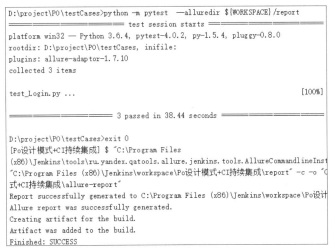

图 2.74　控制台输出界面

图 2.75　Allure 测试报告汇总界面

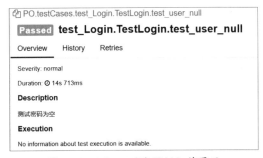

图 2.76　Allure 测试用例细节界面

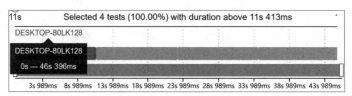

图 2.77　Allure 测试用例执行消耗时间界面

整个自动化测试框架的设计图如图 2.78 所示。

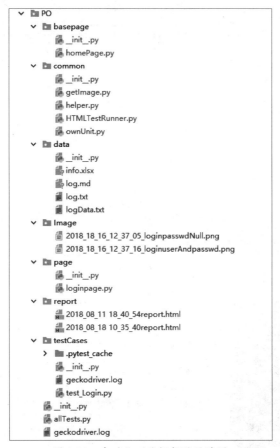

图 2.78　自动化测试框架的设计图

2.16　自动化测试扩展应用实战

有时做自动化测试时并不想弹出浏览器界面，这种情况就需要设置浏览器的无头模式运行方式。当然，为了提高测试效率或对系统进行兼容性测试，也可以结合 Python 多线程技术调用多个浏览器并行执行测试用例。本节只是给读者提供一个测试思路，更多的实际应用还需在项目中不断总结和优化。

2.16.1 配置 Firefox 无头模式

Firefox 无头模式配置非常简单，只需要在自动化测试代码中增加几行代码就可以完成无头模式运行，示例如下：

```
from selenium import webdriver
# 创建新实例
options = webdriver.FirefoxOptions()
# 设置 Firefox 无头模式
options.add_argument('--headless')
options.add_argument('--disable-gpu')
executable_path = r'C:\Python36\geckodriver.exe'   # Firefox 浏览器驱动所在路径
driver = webdriver.Firefox(firefox_options=options,executable_path=executable_path)
```

2.16.2 配置 Chrome 无头模式

Chrome 无头模式的设置与 Firefox 无头模式的设置基本一致，也需要增加 chrome_options.add_argument('--headless') 和 chrome_options.add_argument('--disable-gpu') 两部分代码，示例如下：

```
from selenium import webdriver
# 创建新实例
chrome_options = Options()
# 设置 Chrome 无头模式
chrome_options.add_argument('--headless')
chrome_options.add_argument('--disable-gpu')
driver = webdriver.Chrome(chrome_options=options,options=chrome_options)
```

注意：在 Mac 和 Windows 操作系统下，Chome 59 以上版本对应 Chromedriver 才支持无头模式。

拓展：无头浏览器是一种很好的工具，用于自动化测试和不需要可视化用户界面的服务器。它将 Chromium 和 Blink 渲染引擎提供的所有现代网页平台的特征都转化成了命令行来运行。

2.16.3 多线程调用浏览器运行实战

在无头模式下，使用 Python 多线程调用多个浏览器运行实例，示例如下：

```
from selenium import webdriver
from time import sleep
import threading    # 导入多线程模块
from selenium import webdriver
```

```python
from selenium.webdriver.chrome.options import Options
def test_baidu_search(browser,url):
    if browser == "FireFox":
        # 创建的新实例驱动
        options = webdriver.FirefoxOptions()
        # Firefox 无头模式
        options.add_argument('--headless')
        options.add_argument('--disable-gpu')
        executable_path = r'C:\Python36\geckodriver.exe'
        driver = webdriver.Firefox(firefox_options=options,
                                    executable_path=executable_path)
        # driver = webdriver.Firefox()
    elif browser == "Chrome":
        # 创建的新实例驱动
        chrome_options = Options()
        # Chrome 无头模式
        chrome_options.add_argument('--headless')
        chrome_options.add_argument('--disable-gpu')
        driver = webdriver.Chrome(chrome_options=chrome_options)
        # driver = webdriver.Chrome()
    elif browser == 'Ie':
        driver = webdriver.Ie()
    # 搜索脚本
    driver.get(url)
    sleep(3)
    driver.find_element_by_id("kw").send_keys(u"多线程启动不同浏览器")
    driver.find_element_by_id("su").click()
    sleep(3)
    driver.quit()
```

test_baidu_search() 方法用于判断入参的浏览器类型，如果是 FireFox 浏览器，则调用 FireFox 无头模式运行；如果是 Chrome 浏览器，则调用 Chrome 无头模式运行。无头模式运行状态完成的是百度搜索关键字的场景，示例如下：

```python
if __name__ == "__main__":
    data = {"FireFox": "http://www.baidu.com","Chrome": "http://www.baidu.com",
            "Ie":"http://www.baidu.com"}
    # 构建线程
    threads = []
    for browser,url in data.items():
        t = threading.Thread(target=test_baidu_search,args=(browser,url))
        threads.append(t)
    # 启动所有线程
    for thr in threads:
        thr.start()
```

for 循环遍历 data 字典中所装载的线程并依次将线程追加到 threads 中。start() 方法用于启动所有线程活动。

2.16.4　搭建 PyCharm IDE 开发环境

"工欲善其事，必先利其器"。在编写脚本时选择一款合适的代码编辑器十分必要。目前市场上编辑器众多，如 PyCharm、Sublimetext、Atom、Eclipse、Vim 和 Visual Studio Code 等，这里推荐使用 PyCcharm。

PyCharm 是一种 Python IDE，带有一整套可以帮助用户在使用 Python 语言开发时提高其效率的工具，如调试、语法高亮、Project 管理、代码跳转、智能提示、自动完成、单元测试和版本控制。此外，PyCharm IDE 提供了一些高级功能，以支持 Django 框架下的专业 Web 开发。

1．下载安装

PyCharm 的下载地址为 https://www.jetbrains.com/pycharm/download/，选择 Community 社区版本直接下载，如图 2.79 所示。

图 2.79　PyCharm 版本下载界面

指定一个安装目录，如图 2.80 所示，可将 PyCharm 安装在 D 盘中。

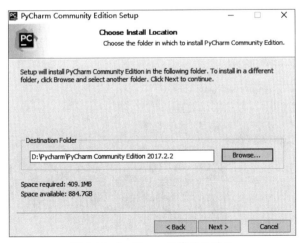

图 2.80　PyCharm 安装目录界面

选中 "I do not have a previous version of PyCharm or I do not want to import my settings" 单选按钮，

如图 2.81 所示。

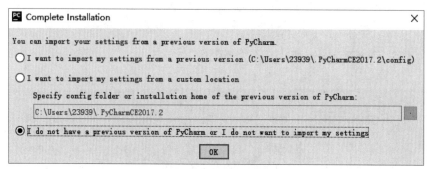

图 2.81　PyCharm 设置首页

PyCharm 背景可以先不修改，后面统一修改，单击"Skip"按钮跳过，如图 2.82 所示。

图 2.82　PyCharm 背景设置界面

2．创建项目工程

- Location：填写本地项目工程路径，可以自定义一个目录位置。
- Interpreter：添加 Python 解释器。一般安装完 Python 环境，这个位置不需要设置，采用默认设置即可。

PyCharm 首页如图 2.83 所示。

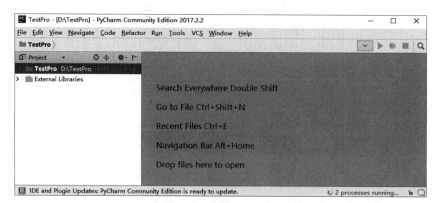

图 2.83　PyCharm 首页

3．设置 PyCharm 字体

选择"File"→"Settings"→"Editor"→"Font"选项，设置后单击"OK"按钮确认即可，如图 2.84 所示。

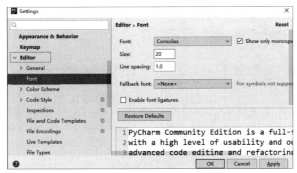

图 2.84　字体设置界面

- Font：字体类型，其下拉列表框中有很多字体可供选择。
- Size：字体大小，可以手动修改。
- Line spacing：行间隔，可以手动修改。

4．调整颜色背景

选择"File"→"Settings"→"Appearance&Behavior"→"Appearance"选项，设置后单击"OK"按钮确认即可，如图 2.85 所示。

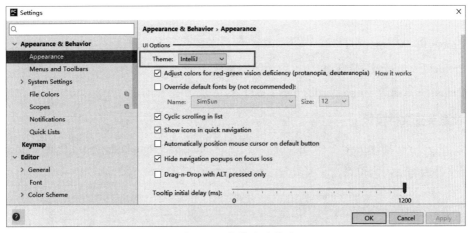

图 2.85　背景设置界面

- Theme：主题。
- Darcula：黑暗背景。

5．去除下划线

选择"File"→"Settings"→"Editor"→"Color Scheme"→"General"选项，取消选中

"Effects"复选框,即可去除下划线,如图2.86所示。

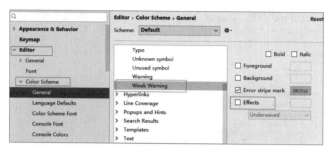

图 2.86　去除下划线界面

6．设置中文编码

选择"File"→"Settings"→"Editor"→"Code Style"→"File Encodings"选项,如图2.87所示。

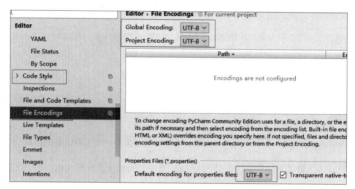

图 2.87　设置中文编码界面

- Global Encodeing：UTF-8。
- Project Encodeing：UTF-8。
- Default encoding for properties files：UTF-8。

7．设置头部文件模板

选择"File"→"Settings"→"Editor"→"File and Code Templates"→"Python Script"选项,配置头部文件内容,内容可以自定义设置,如图2.88所示。

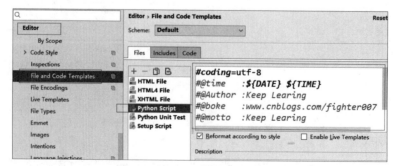

图 2.88　设置头部文件模板界面

第3章 Robot Framework自动化测试实战

Robot Framework 是一款基于 Python 编写的功能自动化测试框架，具备良好的可扩展性，支持关键字驱动，可以同时测试多种类型的客户端或接口，也可以进行分布式测试执行，主要用于轮次很多的验收测试和验收测试驱动开发。

在众多自动化测试工具中，笔者认为 Robot Framework 的使用非常灵活，支持功能自动化测试、接口测试、数据库测试和移动端测试等。如果不想写过多的代码来进行自动化测试，那么 Robot Framework 绝对是不二之选。本章进行 Robot Framework 自动化测试工具的学习。

3.1 搭建 Robot Framework 环境

在使用 Robot Framework 自动化测试工具前，先进行 Robot Framework 自动化测试环境的搭建。

3.1.1 安装 Python

早期版本 RIDE 界面是基于 Python 2 开发的，所以需要安装 Python 2 的环境。关于 Python 2 的环境安装，笔者不再阐述，读者可以去 Python 官网（https://www.python.org/downloads）下载与安装。RIDE 是 Robot Framework 的官方编辑器，可以使用 RIDE 工具来进行测试用例的创建、运行和测试项目的组织等工作。

3.1.2 安装 wxPython

Robot Framework 从 3.0 版本开始支持 Python 3，笔者基于 Windows 平台下载了最新的 RIDE 试用了一下，最大的感受是新版本 RIDE 界面变得漂亮了些，但是表格的易用性比老版本的 RIDE 差一些，使用起来感觉对用户不是很友好。所以本小节依然使用基于 Python 2 开发的 RIDE 界面给读者进行案例演示。需要使用最新的 RIED 界面的读者可以自行上网查看最新版本的 Robot Framework 环境安装教程，安装过程非常得简单。安装完成后，也可以使用最新版本的 RIDE 操作本章案例，是完全兼容的。wxPython 安装界面如图 3.1 所示。

图 3.1 wxPython 安装界面

3.1.3 安装 Robot Framework

Robot Framework 的安装方式分为在线和离线。如果使用在线安装，可以打开 cmd 命令提示符界面，输入"pip install robotframework"，即可完成安装，示例如下：

```
C:\Users\Administrator>pip install robotframework
Collecting robotframework
  Downloading https://files.pythonhosted.org/packages/36/c6/6f89c80ac5a526a091bd
383ffdfc64c9a68d9df0c775d4b36f03d8e0ac25/robotframework-3.1.1-py2.py3-none-any.whl
(601kB)
     100% |████████████████████████████████| 604kB 636kB/s
Installing collected packages: robotframework
Successfully installed robotframework-3.1.1
```

如果使用离线安装，直接从网上下载最新的 Robot Framework 安装包，然后进入 setup.py 所在的文件目录，输入"python setup.py install"命令，即可完成安装。

注意：使用在线安装方式时，一定要确保网络环境是良好的，否则会下载失败。

3.1.4 安装 RIDE

笔者推荐使用 pip 命令在线安装 RIDE。打开 cmd 命令提示符界面，输入"pip install robotframework-ride==1.5.2.1"进行在线安装（建议安装 1.5.2.1 版本），示例如下：

```
C:\Users\Administrator>pip install robotframework-ride==1.5.2.1
Collecting robotframework-ride==1.5.2.1
  Using cached https://files.pythonhosted.org/packages/3c/14/a5f97f5cf5e981f01e8
c0b4c405b0dfc9bc86500cabb044d2c462f73004a/robotframework-ride-1.5.2.1.tar.gz
Installing collected packages: robotframework-ride
  Running setup.py install for robotframework-ride ... done
Successfully installed robotframework-ride-1.5.2.1on->robotframework-ride)
Requirement already satisfied: Pillow in c:\python27\lib\site-packages (from wxP
ython->robotframework-ride)
Requirement already satisfied: typing in c:\python27\lib\site-packages (from Pyp
```

```
ubsub->robotframework-ride)
Installing collected packages: robotframework-ride
Successfully installed robotframework-ride-1.7.3.1
```

3.1.5 验证测试环境

进入目录 C:\Python27\Scripts，找到 ride.py 文件，如图 3.2 所示。

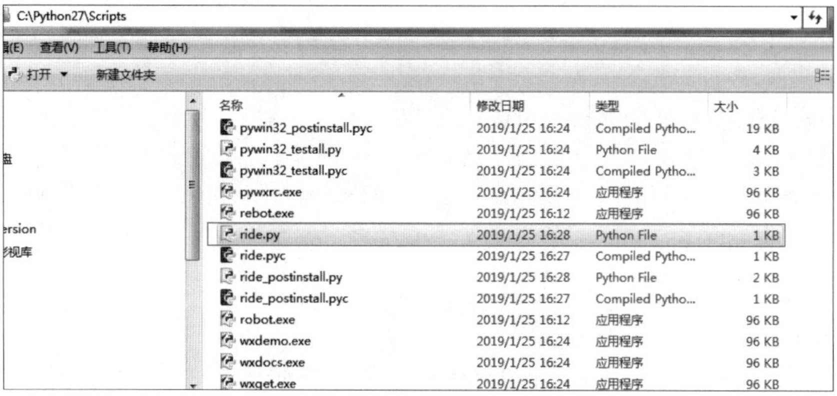

图 3.2　ride.py 文件所在目录界面

双击 ride.py 文件，直接打开 RIDE 界面，选择"File"→"New Project"选项，在弹出的界面中设置好相关选项后，单击"OK"按钮，即可创建一个 New Project（测试项目），如图 3.3 所示。

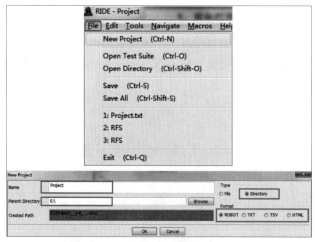

图 3.3　创建测试项目界面

Project 为测试项目名称，类型可以是目录或文件，格式可以用 ROBOT、TXT、TSV 或 HTML，建议选择目录和 TXT 以便于管理。在 Project 目录下右击，在弹出的快捷菜单中选择"New Suite"选项，在弹出的界面中设置好相关选项后，单击"OK"按钮，即可新增一个 New Suite（测试套件）用来管理测试用例文件，如图 3.4 所示。

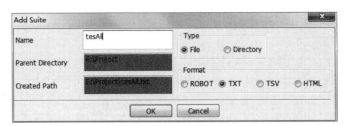

图 3.4　测试套件界面

3.1.6　制作 RIDE 快捷图标

为了更方便地打开 RIDE 工具，在桌面右击，在弹出的快捷中选择"新建"→"快捷方式"选项，在弹出的界面中将 C:\Python27\pythonw.exe -c "from robotide import main; main()" 这段代码输入"请键入对象的位置"下的文本框中，其中 C:\Python27 指的是 Python 的安装目录，如图 3.5 所示。

图 3.5　RIDE 快捷键设置界面

单击"下一步"按钮，给快捷方式命名，如图 3.6 所示，然后单击"完成"按钮。

图 3.6　快捷方式命名界面

后面可以直接通过创建好的快捷方式图标来启动 RIDE 工具界面。

安装与导入 Selenium2Library 库

本节学习 Robot Framework 框架中的 Selenium2Library 库，该库提供了众多底层的关键字来支持 Web 端功能自动化测试。

3.2.1 安装 Selenium2Library 库

笔者推荐使用 pip 命令在线安装 Selenium2Library 库。打开 cmd 命令提示符界面,输入"pip install robotframework-Selenium2Library"进行在线安装,示例如下:

```
C:\Users\Administrator>pip install robotframework-Selenium2Library
Collecting robotframework-Selenium2Library
  Using cached https://files.pythonhosted.org/packages/1c/f1/612f9aa29f33b25a103
4749dde67dfbf6de9b297709d06df71f9bfabfc81/robotframework_selenium2library-3.0.0-
py2.py3-none-any.whl
  Downloading https://files.pythonhosted.org/packages/ff/15/6961c801eeec7f062973
509958b33f158bbc505d45ee6c20b2966275ef51/robotframework_seleniumlibrary-3.3.1-
py2.py3-none-any.whl (81kB)
    50% |████████████████                | 40kB 121kB/s eta 0:00
    62% |████████████████████            | 51kB 144kB/s eta
    75% |████████████████████████        | 61kB 166kB/s
    87% |████████████████████████████    | 71kB 188k
    100% |████████████████████████████████| 81kB
202kB/s
Collecting selenium>=3.4.0 (from robotframework-seleniumlibrary>=3.0.0->
robotfra mework- Selenium2Library)
Successfully installed robotframework-Selenium2Library-3.0.0 robotframework-
sele niumlibrary- 3.3.1 selenium-3.141.0 urllib3-1.24.1
```

3.2.2 导入 Selenium2Library 库

在使用任何一个库下的关键字来实现某个操作时,都需要先将库导入 Robot Framework 中。单击 testALL 测试套件下的"Library"按钮,在弹出界面的"Name"文本框中输入"Selenium2Library", 单击"OK"按钮,即可将 Selenium2Library 库导入 Robot Framework 中,如图 3.7 所示。

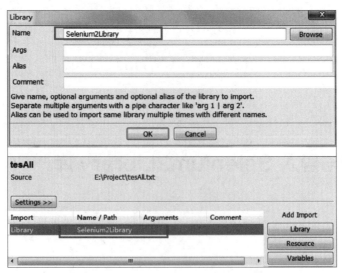

图 3.7 导入 Selenium2Library 库界面

这里需要说明的是，导入库时一定要注意大小写及注意检查库的名称是否正确。如果导入的库显示为红色，表示导入的库不存在；如果是黑色则表示导入成功。

按 F5 键，查看 Selenium2Library 库的关键字说明，如图 3.8 所示。

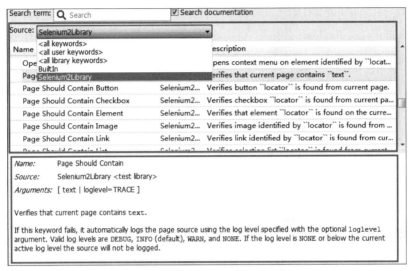

图 3.8　Selenium2Library 库的关键字说明

帮助说明分为上下两个部分，上半部分表示库中的所有关键字清单，下半部分表示库中的指定关键字的用法示例。

 浏览器驱动配置

开始自动化测试前，需要先将浏览器驱动配置好。配置策略要参考浏览器和浏览器版本之间的驱动关系。

3.3.1　配置 Firefox 浏览器驱动

如果使用 Firefox 浏览器进行 Web 自动化测试，需要知道 Selenium 2.xx 版本是否自动集成了 Firefox 驱动，只需要对应 Selenium 版本与 Firefox 浏览器版本即可。Selenium 版本和 Firefox 浏览器参考对照表如图 3.9 所示。

Selenium 3.xx 的基本要求如下。

Selenium版本	Firefox浏览器
2.25.0	V18
2.30.0	V19
2.31.0	V20
2.42.2	V29
2.44.0	V33(不支持31)
2.52.0	V45.0
2.53.0	V46.0
2.53.1	V47.0.1

图 3.9　Selenium 版本和 Firefox 浏览器参考对照表

（1）从 Selenium 3.0.0 开始就要求 Firefox 为 48 及以上版本。

（2）Selenium 3.x 使用的 Java 版本为 JDK 1.8。

（3）Selenium 3.x 使用 geckodriver 作为 Firefox 浏览器驱动的替代。

Selenium 2.0 与 Selenium 3.0 的主要区别如下。

（1）Selenium 2 启动 Firefox 浏览器不需要 geckodriver 驱动。

（2）Selenium 3 需要的 Java 最低版本是 Java 8。

（3）Selenium 3 启动 Firefox 浏览器也需要像其他浏览器一样安装驱动，驱动名为 geckodriver。

（4）Selenium 3 核心的安装包中彻底删除了 Selenium RC。

（5）Selenium 3 如果要运行从 IDE 转化过来的用例，需要单独安装 leg-rc 的安装包。

（6）Selenium 3 支持在 MacOS 上使用 Safari。

（7）Selenium 3 支持微软最新的浏览器 Edge。

3.3.2 配置 Chrome 浏览器驱动

使用 Chrome 浏览器进行功能自动化测试时，要知道 Chrome 浏览器版本和驱动的对应关系。Chrome driver 版本和 Chrome 浏览器参考对照表如图 3.10 所示。

Chrome driver 版本	Chrome 浏览器
V2.40	v66-68
V2.39	v66-68
V2.38	v65-67
V2.37	v64-66
V2.36	v63-65
V2.35	v62-64
V2.34	v61-63
V2.33	v60-62
V2.32	v59-61
v2.31	v58-60
v2.30	v58-60
v2.29	v56-58
v2.28	v55-57
v2.27	v54-56
v2.26	v53-55
v2.25	v53-55

图 3.10　Chrome driver 版本和 Chrome 浏览器参考对照表

Chrome 驱动的下载地址为 http://chromedriver.storage.googleapis.com/index.html。下载好驱动后，需要将 Chrome 驱动复制到 Python 的安装目录 C:\Python27，或者将驱动所在的路径追加到系统的环境变量 PATH 中。

3.4 元素定位实战

元素定位在自动化测试实战中非常重要,本节演示 Selenium2Library 库中关于元素定位的实战。

3.4.1 引用 id

如果页面中的某个元素存在 id 属性,则可以直接用 id 来定位,这种方式也更加具体。在 testAll 测试套件中右击,在弹出的快捷菜单中选择"New TestCase"选项,新建一个名为"id 定位实战"的测试用例,案例如图 3.11 所示。

#打开浏览器		
Open Browser	http://www.baidu.com	gc
#通过id定位百度文本框		
Input text	id=kw	python
#单击百度一下按钮		
Click Element	id=su	
#打印搜索成功		
Log	搜索成功!	
#关闭浏览器		
Close Browser		

图 3.11 id 定位案例

- Open Browser:用于打开百度浏览器。http://www.baidu.com 表示测试地址,gc 表示是通过 Chrome 浏览器打开测试地址的。
- Input text:表示向文本框输入内容。id=kw 表示百度文本框的 id 属性值是 kw;Python 是参数,即搜索的内容。
- Click Element:表示单击元素。id=su 表示百度按钮元素本身的 id 属性。
- Log:相当于 Python 中的 Print 输出功能,表示输出"搜索成功"。
- Close Browser:用于关闭浏览器。

按 F8 键,查看 RIDE 日志的输出结果:

```
Starting test: Project.tesAll.id定位实例演示
20190127 12:50:11.240 : INFO : Opening browser 'gc' to base url 'http://www.baidu.com'.
20190127 12:50:14.687 : INFO : Typing text 'python' into text field 'id=kw'.
20190127 12:50:14.885 : INFO : Clicking element 'id=su'.
20190127 12:50:14.988 : INFO : 搜索成功!
Ending test:   Project.tesAll.id定位实例演示
```

特殊情况:一般来说,库中的关键字不区分大小写,但为了编写规范,建议首字母为大写。在使用 id 进行定位时,可以忽略 id 标识,直接写 id 对应的值即可。在 Robot Framework 库中显示为蓝色的字体都是关键字。

如果按 F8 键后，控制台出现如下报错信息：

```
command: pybot.bat --argumentfile c:\users\admini~1\appdata\local\temp\RIDEtlkg3c.d\argfile.txt --listenerC:\Python27\lib\site-packages\robotide\contrib\testrunner\TestRunnerAgent.py:55374:False E:\Project [Error 2]
```

则解决办法是检查 Python 安装目录 C:\Python27\Scripts 下是否存在 pybot.bat 文件，如果不存在则新建一个 pybot.bat 文件，并填入如下代码：

```
@echo off
python -m robot.run %*
```

再次运行即可成功。

更多驱动及浏览器简写参考对照表如图 3.12 所示。

关键字	浏览器/设备
firefox	Firefox
ff	Firefox
internetexplorer	Internet Explorer
ie	Internet Explorer
googlechrome	Google Chrome
gc	Google Chrome
chrome	Google Chrome
opera	Opera
phantomjs	PhantomJS
htmlunit	HTMLUnit
htmlunitwithjs	HTMLUnit with Javascript support
android	Android
iphone	Iphone
safari	Safari
edge	Edge

图 3.12　各驱动及浏览器简写参考对照表

3.4.2　引用 name

name 定位可以识别第一个匹配名称属性的 UI 元素。name 定位案例如图 3.13 所示。

#打开浏览器		
Open browser	http://www.baidu.com	gc
#通过name定位百度文本框		
Input text	name=wd	python
#单击百度一下按钮		
Click Element	su	
Log	搜索成功！	
#关闭浏览器		
Close Browser		

图 3.13　name 定位案例

按 F8 键，查看 RIDE 日志的输出结果：

```
Starting test: Project.tesAll.name 定位案例
20190128 11:56:54.157 :   INFO : Opening browser 'gc' to base url 'http://www.baidu.com'.
20190128 11:56:57.679 :   INFO : Typing text 'python' into text field 'name=wd'.
20190128 11:56:57.918 :   INFO : Clicking element 'su'.
20190128 11:56:58.053 :   INFO : 搜索成功!
Ending test:   Project.tesAll.name 定位案例
```

注意：id、name 定位器使得 Selenium 可以不依赖于 UI 元素在页面上的位置而进行测试。所以，当页面结构发生变化时，测试依然可以通过。有时设计人员会频繁改动页面，此时通过 id 和 name 特征定位元素就变得非常重要。所以，建议在用元素定位时优先考虑 id、name 方式，如果无法满足再考虑其他定位方式。

3.4.3 引用 link

link 定位是指超链接定位，可以通过链接文字来定位超链接。如果两个链接文字相同，那么第一个匹配的将被识别出来。

这种定位方式比其他方式更便于定位超链接，但需要注意的是，超链接并不是按钮，不能使用关键字 Click Button，而应该使用单击超链接的关键字 Click Link。

link 定位案例如图 3.14 所示。

#打开浏览器		
Open browser	http://www.baidu.com	gc
Sleep	3	
#通过link定位新闻链接		
Click Link	link=新闻	
Log	单击成功!	
#关闭浏览器		
Close Browser		

图 3.14　link 定位案例

- link= 新闻：表示定位文本链接为新闻的元素。
- Sleep：表示休眠为 3s。

按 F8 键，查看 RIDE 日志的输出结果：

```
Starting test: Project.tesAll.link
20190127 16:40:31.086 :   INFO : Opening browser 'gc' to base url 'http://www.baidu.com'.
20190127 16:40:37.974 :   INFO : Slept 3 seconds
20190127 16:40:37.977 :   INFO : Clicking link 'link=新闻'.
20190127 16:40:39.495 :   INFO : 单击成功!
Ending test:   Project.tesAll.link
```

3.4.4 引用 css

css 定位语法更加强大和灵活，也比较简洁。css 定位案例如图 3.15 所示。

#打开浏览器		
Open Browser	http://www.baidu.com	gc
#休眠3s		
Sleep	3	
#向百度文本框输入python		
Input text	css=.s_ipt	python
#单击百度一下按钮		
Click Element	css=#su	
#关闭浏览器		
Close Browser		

图 3.15　css 定位案例

css 语法中的 class 属性统一用"."来标识，id 属性定位用"#"标识。另外，也可以使用 css 属性定位，如 css=input[id='kw']，其中 input 表示定位元素本身的标签名，id 是元素本身的属性，kw 是属性对应的属性值。

此外，也可以采用 css 层级与属性结合定位元素。例如，定位百度文本框可以写成 css=.quickdelete-wrap>input#kw，其中 .quickdelete-wrap 表示文本框父类元素本身 class 的属性值，> 表示父类标签的下一级元素，如图 3.16 所示。

#向百度文本框输入python		
Input Text	css=.quickdelete-wrap>input#kw	python

图 3.16　css 层级与属性结合定位案例

如果父类无法定位元素，这时可以找到父类的上一级元素。定位百度文本框还可以写成 css=form#form>span>input，如图 3.17 所示。

#向百度文本框输入python		
Input Text	css=form#form>span>input	python

图 3.17　父子关系定位案例

按 F8 键，查看 RIDE 日志的输出结果：

```
Starting test: Project.tesAll.css1
20190127 17:36:39.645 : INFO : Opening browser 'gc' to base url 'http://www.baidu.com'.
20190127 17:36:46.483 : INFO : Slept 3 seconds
20190127 17:36:46.483 : INFO : Typing text 'python' into text field 'css=.s_ipt'.
20190127 17:36:46.721 : INFO : Clicking element 'css=#su'.
20190127 17:36:50.064 : INFO : Typing text 'python' into text field 'css=.quickdelete-wrap>input#kw'.
20190127 17:36:50.064 : INFO : Cannot capture screenshot because no browser is open.
20190127 17:36:50.080 : FAIL : No browser is open.
Ending test:    Project.tesAll.css1
```

3.4.5　引用 xpath

XPath 是 XML 和 Path 的缩写，主要用于在 XML 文档中选择文档中的节点。XPath 语言可以用在 HTML 中寻找指定的节点。xpath 定位与 css 定位相比有更大的灵活性，但 xpath 的定位速度比 css 慢。下面看一个 xpath 定位案例，如图 3.18 所示。

#打开浏览器		
Open browser	http://www.baidu.com	gc
#休眠3s		
Sleep	3	
#通过xpath属性定位文本框		
Input text	//input[@id='kw']	python
#单击百度一下按钮		
Click element	//input[@id='su']	

图 3.18　xpath 定位案例

//input[@id='kw']：表示使用 xpath 相对路径来定位，其中 // 表示相对路径；@id='kw' 表示定位 id 属性值为 kw 的元素，@ 后面可以跟任意一个具有唯一性的属性。

此外，也可以采用 xpath 层级与属性结合定位元素。例如，定位百度文本框也可以写成 //span[@class="bg s_ipt_wr quickdelete-wrap"]/input 或 //form[@id='form']/span/input，如图 3.19 所示。

#通过xpath父类属性定位文本框1		
Input Text	//span[@class="bg s_ipt_wr quickdelete-wrap"]/input	python
#通过xpath父类属性定位文本框2		
Input Text	//form[@id='form']/span/input	python

图 3.19　xpath 层级与属性结合定位案例

也可以使用逻辑运算符 and 或 or 来连接元素的某两个属性，加强元素的唯一性，如图 3.20 所示。

#通过逻辑运算符and定位百度文本框		
Input Text	//input[@id='kw' and @name='wd']	python
#通过逻辑运算符and定位百度按钮		
Click Element	//input[@id='su' and @type='submit']	

图 3.20　xpath 逻辑运算符 and 定位案例

按 F8 键，查看 RIDE 日志的输出结果：

```
Starting test: Project.tesAll.xpath
20190127 17:20:15.269 : INFO : Opening browser 'gc' to base url 'http://www.baidu.com'.
20190127 17:20:21.284 : INFO : Slept 3 seconds
20190127 17:20:21.284 : INFO : Typing text 'python' into text field '//input[@id='kw']'.
20190127 17:20:21.489 : INFO : Clicking element '//input[@id='su']'.
Ending test:   Project.tesAll.xpath
```

3.4.6　xpath 定位动态属性

有些情况下，定位的元素属性值是动态变化的，每刷新一次就会变化一次。这种情况除可以使用 xpath 层级路径来定位外，还可以使用 xpath 提供的 contains() 函数来定位，如图 3.21 所示。

图 3.21　iframe 标签下的 id 属性界面

如图 3.21 所示，iframe 标签下的 id 属性值每刷新一次就会变化一次。使用 xpath 的 contains() 函数来定位，案例如图 3.22 所示。

#打开126首页		
Open Browser	https://mail.126.com/	gc
#切换iframe标签		
Select Frame	//iframe[starts-with(@id,'x-URS-iframe')]	
#输入账号		
Input Text	name=email	IT测试老兵

图 3.22　contains() 函数定位动态值案例

- Select Frame：表示切换 iframe 标签需要用到的关键字。因为要定位的元素（账号文本框）处在 iframe 标签中，所以必须要先切换到 iframe 标签中才可以操作元素。
- //iframe[starts-with(@id,'x-URS-iframe')]：表示在 HTML 页面中查找 iframe 标签下 id 的属性值以 x-URS-iframe 开头的元素。此外，contains() 函数还有 ends-with() 方法，它的用法和 starts-with 一样。

3.5　JQuery 定位实战

JQuery 选择器支持对 HTML 元素组或单个元素进行操作。它基于已经存在的 css 选择器，支持通过 id、class 类型、属性、属性值等方法来定位 HTML 元素。在自动化测试实战中，会经常使用 JQuery 来调试并且定位页面中的元素。

3.5.1 处理特殊单击事件

有些情况下，使用 Selenium2Library 库中的元素定位方法定位某个元素时，单击事件不生效。此时可以借助 JQuery 来实现单击事件。进入猫眼电影首页，按 F12 键，在 Console 调试器界面输入 JQuery 代码，如图 3.23 所示。

图 3.23　Console 调试界面使用 JQuery 模拟单击事件

图 3.23 中，$('img.ranking-img.default-img').click() 表示模拟单击页面上的某一部电影操作。

除使用 JQuery 模拟单击事件外，还可以使用 JavaScript 来操作，如图 3.24 所示。

图 3.24　Console 调试界面使用 JavaScript 模拟单击事件

使用 JQuery 模拟单击事件的案例如图 3.25 所示。

#打开浏览器		
Open Browser	https://maoyan.com/	gc
Sleep	3	
#通过 JQuery 模拟单击事件		
Execute JavaScript	$("img.ranking-img.default-img").click()	
Sleep	3	
Close Browser		

图 3.25　JQuery 单击事件案例

Exectue JavaScript：执行 JavaScript 或 JQuery 语法的关键字。

按 F8 键，查看 RIDE 日志的输出结果：

```
Starting test: Project.tesAll.jquery
20190127 19:18:46.350 : INFO : Opening browser 'gc' to base url 'https://maoyan.com/'.
20190127 19:18:53.871 : INFO : Slept 3 seconds
20190127 19:18:53.873 : INFO :
Executing JavaScript:
$('img.ranking-img.default-img').click()
Without any arguments.
20190127 19:18:57.117 : INFO : Slept 3 seconds
Ending test:   Project.tesAll.jquery
```

3.5.2 移除 readOnly 属性

在做自动化测试时，有时会遇到在输入日期时无法输入的情况。这是因为输入框是 readOnly 属性，需要对输入框进行处理，即移除它的 readOnly 属性，这样即可正常输入日期，如图 3.26 所示。

图 3.26　移除 readOnly 属性界面

移除 readOnly 属性案例如图 3.27 所示。

#打开12306网站		
Open Browser	https://kyfw.12306.cn/otn/leftTicket/init	gc
Sleep	3	
#移除出发日期readOnly属性		
Execute JavaScript	$('#train_date').removeAttr('readonly')	
#输入出发日期为2019-09-25		
Input Text	id=train_date	2019-09-25
#关闭浏览器		
Close Browser		

图 3.27　移除 readOnly 属性案例

按 F8 键，查看 RIDE 日志的输出结果：

```
Starting test: Project.tesAll.data
20190925 19:34:17.047 : INFO : Opening browser 'gc' to base url 'https://kyfw.12306.cn/otn/leftTicket/init'.
20190925 19:34:24.078 : INFO : Slept 3 seconds
20190925 19:34:24.084 : INFO :
Executing JavaScript:
$('# train_date').removeAttr('readonly')
Without any arguments.
20190925 19:34:24.103 : INFO : Typing text '2019-09-25' into text field 'id=train_date'.
Ending test: Project.tesAll.data
```

3.5.3 处理 Display 隐藏元素

有些情况下，定位的元素被隐藏了，这时可以使用 JQuery 来处理隐藏元素，如图 3.28 所示。

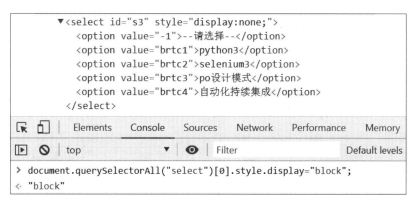

图 3.28　使用 JQuery 处理隐藏元素界面

使用 JavaScript 语法处理隐藏的元素内容，案例如图 3.29 所示。

#打开测试地址		
Open Browser	file:///E:/display.html	gc
Sleep	3	
#JQquery设置dispaly=none		
Execute JavaScript	document.querySelectorAll("select")[0].style.display="block";	
#打印设置成功		
Log	设置成功	

图 3.29　使用 JavaScript 处理隐藏元素案例

document.querySelectorAll("select")：表示选择所有的 select 标签，[0] 指定元素是这一组标签中的第几个，style.display="block"; 表示设置样式 display="block" 为可见。

此外，JQuery 还提供了其他方法来设置元素的显示和隐藏。

- $("#id").show()：表示为 display:block。
- $("#id").hide()：表示为 display:none。
- $("#id").css('display', 'none')：表示将元素设置为隐藏。
- $("#id").css('display', 'block')：表示将元素设置为显示。

按 F8 键，查看 RIDE 日志的输出结果：

```
Starting test: Project.tesAll.dispaly
20190128 10:17:44.845 : INFO : Opening browser 'gc' to base url 'file:///E:/display.html'.
20190128 10:17:50.808 : INFO : Slept 3 seconds
20190128 10:17:50.815 : INFO :
Executing JavaScript:
document.querySelectorAll("select")[0].style.display="block";
Without any arguments.
20190128 10:17:50.840 : INFO : 设置成功
Ending test:   Project.tesAll.dispaly
```

3.5.4 使用 JQuery 处理蒙层事件

图 3.30 所示的区域有一蒙层，蒙层在写用例时干扰比较大，以至于无法对蒙层下的元素进行操作，即使操作了也会提示该操作的元素不能被单击。单击事件被另一个元素接收到了，其实指的就是页面中的蒙层。

图 3.30　蒙层事件界面

通过 JQuery 提供的 $("#id").hide() 方法可以隐藏图片，去掉蒙层，其中 () 内支持 css 选择器中的 id 和 class 选择器。

3.5.5 使用 JQuery 获取文本框中的值

使用 JQuery 可以获取文本框中的值。进入百度首页，按 F12 键，查看 Console 调试界面，如图 3.31 所示。

图 3.31　使用 JQuery 获取文本框中的值调试界面

图 3.31 中，#kw 表示文本框的 id 属性值是 kw，val() 方法用于获取文本框中的值。

3.5.6 使用 JQuery 向文本框中输入内容

使用 JQuery 向文本框中输入内容，只需要在 val() 方法中写入参数即可，如图 3.32 所示。

图 3.32　使用 JQuery 向文本框中输入内容

3.6　获取窗口标题

在自动化测试过程中，窗口标题经常被用来作为断言信息，从而验证测试结果。获取窗口标题案例如图 3.33 所示。

#打开浏览器		
Open Browser	http://www.baidu.com	gc
Sleep	3	
#输入 robotframework		
Input Text	id=kw	robotframework
#单击百度按钮		
Click Element	id=su	
#休眠3秒		
Sleep	3	
#获取窗口标题		
${title}	get title	
Log	${title}	
#关闭浏览器		
Close Window		

图 3.33　获取窗口标题案例

get title：用于获取窗口标题并将标题赋值给 ${title}。在 Robot Framework 中定义变量用 $ 来声明。按 F8 键，查看 RIDE 日志的输出结果：

```
Starting test: Project.tesAll.get
20190128 11:01:29.855 :  INFO : Opening browser 'gc' to base url 'http://www.baidu.com'.
20190128 11:01:36.349 :  INFO : Slept 3 seconds
20190128 11:01:36.351 :  INFO : Typing text 'robotframework' into text field 'id=kw'.
20190128 11:01:36.702 :  INFO : Clicking element 'id=su'.
20190128 11:01:39.837 :  INFO : Slept 3 seconds
20190128 11:01:39.853 :  INFO : ${title} = robotframework_百度搜索
20190128 11:01:39.856 :  INFO : robotframework_百度搜索
Ending test:    Project.tesAll.get
```

3.7 获取文本信息

在 Robot Framework 中可以使用 get text 关键字来获取页面中的文本信息。在实际测试过程中，会经常使用文本信息作为断言从而验证测试结果是否通过，如图 3.34 所示。

#打开浏览器		
Open Browser	http://www.baidu.com	gc
Sleep	3	
#输入 robotframework		
Input Text	id=kw	robotframework
#单击百度按钮		
Click Element	id=su	
#休眠3秒		
Sleep	3	
#获取页面文本信息		
${text}	get text	link=英文结果
Log	${text}	
#关闭浏览器		
Close Window		

图 3.34　获取文本信息案例

使用 get text 关键字获取文本信息时，需要获取页面中文本对应的元素（link= 英文结果），然后赋值给 ${text} 变量，最后输出文本信息。

按 F8 键，查看 RIDE 日志的输出结果：

```
Starting test: Project.tesAll.text
20190128 11:22:04.642 :  INFO : Opening browser 'gc' to base url 'http://www.baidu.com'.
20190128 11:22:11.608 :  INFO : Slept 3 seconds
20190128 11:22:11.608 :  INFO : Typing text 'robotframework' into text field 'id=kw'.
20190128 11:22:11.915 :  INFO : Clicking element 'id=su'.
20190128 11:22:15.037 :  INFO : Slept 3 seconds
20190128 11:22:15.223 :  INFO : ${text} = 英文结果
20190128 11:22:15.233 :  INFO : 英文结果
Ending test:  Project.tesAll.text
```

3.8 鼠标指针悬停实战

在自动化测试过程中，有些菜单是悬浮菜单，需要鼠标指针悬停在这些菜单上时下级元素才会显示。例如，对于这类控件的处理，Selenium2Library 库中提供了 Mouse Over、Mouse Down 和

Mouse Up 等方法来实现鼠标指针悬停操作，案例如图 3.35 所示。

1	Open Browser	http://www.baidu.com	gc
2	sleep	5	
3	#最大化浏览器窗口		
4	Maximize Browser Window		
5	Comment	鼠标悬停	放到百度首页更多产品
6	Mouse Over	tj_briicon	
7	sleep	5	
8	Comment	按住鼠标左键不松开	单击百度糯米 单击不松开
9	Mouse Down	tj_nuomi	
10	sleep	5	
11	Comment	按住鼠标左键后释放	单击百度糯米 单击释放
12	Mouse Up	tj_nuomi	
13	sleep	5	
14	Comment	鼠标悬停	悬停到深圳
15	Mouse Over	css=span.arrow-down-line	
16	sleep	5	
17	#单击北京		
18	Click Element	link=北京	

图 3.35　鼠标指针悬停案例

在上述案例中，Mourse Over 用来模拟鼠标指针悬停到某个元素上，Mouse Down 模拟按住鼠标左键不松开操作，Mouse Up 用来模拟按住鼠标左键后释放操作。

按 F8 键，查看 RIDE 日志的输出结果：

```
Starting test: Project.tesAll.mourse
20190128 11:48:31.639 :  INFO : Opening browser 'gc' to base url 'http://www.baidu.com'.
20190128 11:48:39.742 :  INFO : Slept 5 seconds
20190128 11:48:40.894 :  INFO : Simulating Mouse Over on element 'tj_briicon'.
20190128 11:48:45.989 :  INFO : Slept 5 seconds
20190128 11:48:45.989 :  INFO : Simulating Mouse Down on element 'tj_nuomi'.
20190128 11:48:51.057 :  INFO : Slept 5 seconds
20190128 11:48:51.057 :  INFO : Simulating Mouse Up on element 'tj_nuomi'.
20190128 11:48:58.760 :  INFO : Slept 5 seconds
20190128 11:48:58.760 :  INFO : Simulating Mouse Over on element 'css=span.arrow-down-line'.
20190128 11:49:03.812 :  INFO : Slept 5 seconds
20190128 11:49:03.812 :  INFO : Clicking element 'link=北京'.
Ending test:   Project.tesAll.mourse
```

3.9　操作多窗口实战

当定位的元素不在当前窗口时，需要切换窗口来定位元素，案例如图 3.36 所示。

Open Browser	http://www.baidu.com	gc
Sleep	2	
Input Text	id=kw	渗透吧
Click Element	id=su	
Sleep	3	
#单击渗透吧链接		
Click Element	//*[@id="1"]/h3/a/em	
#切换窗口 title		
Sleep	3	
Select window	title=渗透吧-百度贴吧	
Sleep	3	
Click Element	link=进入贴吧	
Log	加载完毕！！！！	
Sleep	3	
#回退到第一个窗口		
Select window	title=渗透吧_百度搜索	
Reload Page	#刷新页面	
log	切换回首个窗口successfull!!	

图 3.36 多窗口案例

- Select window：表示切换窗口的关键字。
- title= 渗透吧 - 百度贴吧：表示切换到窗口标题为"渗透吧 - 百度贴吧"的窗口。除使用 title 关键字外，还可以使用 new 关键字来实现此功能，如图 3.37 所示。

Select window	new

图 3.37 使用 new 关键字来切换窗口

按 F8 键，查看 RIDE 日志的输出结果：

```
Starting test: Project.tesAll.window
20190128 12:23:43.960 : INFO : Opening browser 'gc' to base url 'http://www.baidu.com'.
20190128 12:23:49.674 : INFO : Slept 2 seconds
20190128 12:23:49.674 : INFO : Typing text '渗透吧' into text field 'id=kw'.
20190128 12:23:49.867 : INFO : Clicking element 'id=su'.
20190128 12:23:52.973 : INFO : Slept 3 seconds
20190128 12:23:52.973 : INFO : Clicking element '//*[@id="1"]/h3/a/em'.
20190128 12:23:56.139 : INFO : Slept 3 seconds
20190128 12:24:00.427 : INFO : Slept 3 seconds
20190128 12:24:00.427 : INFO : Clicking element 'link=进入贴吧'.
20190128 12:24:02.223 : INFO : 加载完毕！！！！
20190128 12:24:05.226 : INFO : Slept 3 seconds
20190128 12:24:05.796 : INFO : 切换回首个窗口successful!!!
Ending test: Project.tesAll.window
```

3.10 操作下拉列表框实战

在实际自动化测试过程中，经常会遇到一些下拉列表框控件，打开下拉列表框后可以看到有很多选项，可手动选择所要测试的选项。Robot Framework 提供了 3 个常用的操作下拉列表框的关键字。

Select 标签下拉列表框案例如图 3.38 所示。

```
▼<select id="s1Id">
    <option></option>
    <option value="o1val">O1</option>
    <option value="o2val">文本定位</option>
    <option value="o3val">o3</option>
    <option value="o4val">o4</option>
    <option value="o4val">o5</option>
</select>
```

图 3.38　Select 标签下拉列表框案例

定位下拉列表框控件案例如图 3.39 所示。

Open Browser	file:///E:/xialakuang.html	gc
Sleep	3	
#通过索引位定位		
Select From List By Index	id=s1Id	0
Sleep	3	
Close All Browsers		

图 3.39　定位下拉列表框控件案例

Select From List By Index：表示通过索引来定位下拉列表框。id=s1Id 表示定位到 Select 标签，0 表示选择下拉列表框中的第一个选项。

此外，Select 还提供了 Lablel（可以理解为文本描述）定位和 Value 属性值定位方法，案例如图 3.40 所示。

#通过文本定位		
Select From List By Label	id=s1Id	文本定位
#通过属性值value定位		
Select From List By Value	id=s1Id	O1

图 3.40　Select 其他定位方法案例

注意：上述案例演示的是 Select 标签实现下拉列表框控件的定位，如果是非 Select 标签，则可直接实现下拉列表框控件的定位。

操作警告框实战

Selenium2Library 库中提供了 Confirm Action 和 Choose Cancel On Next Confirmation 两个关键字来处理警告框，案例如图 3.41 所示。

Open Browser	http://192.168.1.108:4444/ecshop/user.php	gc
Input Text	username	FighterLu
Input Text	password	123456
Click Element	submit	
Click Element	link=GSM手机	
Click Link	link=诺基亚E66	
#单击加入购物车		
Click Element	//*[@id="ECS_FORMBUY"]/ul/li[9]/a[1]/img	
Sleep	4	
#单击删除		
Click Element	link=删除	
Sleep	3	
#接收警告框		
Confirm Action		

图 3.41　操作警告框案例

上述案例中，Confirm Action 关键字用来确认警告框动作。如果取消警告框操作，则可使用 Choose Cancel On Next Confirmation 关键字。

 获取结果断言

在自动化测试过程中，断言是非常重要的，脚本中如果没有断言，就无法证明自动化测试结果是通过还是失败。Robot Framework 提供了非常多的断言方法，本节会列举一些和 Web 自动化测试相关的断言方法。

【例 3-1】Should (Not) Contain 验证参数 1 中（不）包含参数 2，案例如图 3.42 所示。

#验证参数1中是否包含参数2			
@{listA}	Set Variable	99	100
Should Contain	${listA}	99	
#验证参数1中不包含参数2			
Should Not Contain	${listA}	555	

图 3.42　Should (Not) Contain 断言案例

@{listA} 表示创建一个列表，Set Variable 关键字用于创建变量。

分别进行两次结果断言，第一次用来验证 99 在 ${listA} 中，第二次用来验证 555 不在 ${listA} 变量中，两次断言结果都是正确的。需要说明的是，在 Robot Framework 中，所有字符类型都是字符串类型。

按 F8 键，查看 RIDE 日志的输出结果：

```
Starting test: Project.tesAll.assert
20190128 17:35:06.110 : INFO : @{listA} = [ 99 | 100 | 66 ]
Ending test:   Project.tesAll.assert
```

【例 3-2】Should (Not) Be Empty 验证给定条件（不）为空，案例如图 3.43 所示。

#验证给定条件为空		
${empty}	Set Variable	
Should Be Empty	${empty}	
#验证给定条件不为空		
${empty_Not_null}	Set Variable	111
Should Not Be Empty	${empty_Not_null}	

图 3.43　Should (Not) Be Empty 断言案例

Should Be Empty 关键字用来验证定义的 ${empty} 变量应该为空，Should Not Be Empty 关键字用来验证定义的 ${empty_Not_null} 变量应该不为空，两次断言结果都是正确的。

按 F8 键，查看 RIDE 日志的输出结果：

```
Starting test: Project.tesAll.empty
20190128 17:37:09.539 : INFO : ${empty} = 
20190128 17:37:09.539 : INFO : Length is 0
20190128 17:37:09.539 : INFO : ${empty_Not_null} = 111
20190128 17:37:09.539 : INFO : Length is 3
Ending test:   Project.tesAll.empty
```

【例 3-3】Should (Not) Be Equal 验证变量 A（不）等于变量 B，案例如图 3.44 所示。

#验证变量A等于变量B		
${A}	Set Variable	www.cnblogs.com/fighter007
${B}	Set Variable	www.cnblogs.com/fighter007
Should Be Equal	${A}	${B}
#验证变量A不等于变量B		
${not_A}	Set Variable	www.cnblogs.com/fighter007
${not_B}	Set Variable	fighter007
Should Not Be Equal	${A}	${B}

图 3.44　Should (Not) Be Equal 断言案例

Should Be Equal 关键字判断 ${A} 和 ${B} 是否相等，相等则测试结果通过；Should Not Be Equal 关键字用来判断 ${not_A} 和 ${not_B} 是否不相等，如果不相等，则测试结果通过。

按 F8 键，查看 RIDE 日志的输出结果：

```
Starting test: Project.tesAll.assertequal
20190128 17:41:39.247  :  INFO : ${A} = www.cnblogs.com/fighter007
20190128 17:41:39.247  :  INFO : ${B} = www.cnblogs.com/fighter007
20190128 17:41:39.247  :  INFO : ${not_A} = www.cnblogs.com/fighter007
20190128 17:41:39.247  :  INFO : ${not_B} = fighter007
2019012817:41:39.247:FAIL : www.cnblogs.com/fighter007 == www.cnblogs.com/fighter007
Ending test:   Project.tesAll.assertequal
```

【例 3-4】Should (Not) Be True 验证给定条件（不）为 True，案例如图 3.45 所示。

#验证给定条件为True		
${args1}	Set Variable	521
${args2}	Set Variable	10000
Should Be True	${args1}<${args2}	
#验证给定条件不为True		
Should Not Be True	${args1}>${args2}	

图 3.45　Should (Not) Be True 断言案例

Should Be True 关键字用来测试结果为真，${args1}<${args2}，结果显然是成立的，所以测试结果通过；Should Not Be True 关键字用来判断测试结果不为真，${args1}>${args2}，结果显然是成立的，所以测试结果也通过。

【例 3-5】Page Should Contain 断言案例如图 3.46 所示。

#打开浏览器		
Open Browser	http://www.baidu.com	gc
#通过id定位百度文本框		
Input text	id=kw	python
#单击百度一下按钮		
Click Element	id=su	
Sleep	3	
#断言页面应该包含文本		
Page Should Contain	英文结果	
#关闭浏览器		
Close Browser		

图 3.46　Page Should Contain 断言案例

Page Should Contain 关键字用来断言搜索后的结果页面中应该包含英文结果文本。如果包含则测试结果通过，反之，则失败。

按 F8 键，查看 RIDE 日志的输出结果：

```
Starting test: Project.tesAll.Page Should
20190128 17:54:56.139 : INFO : Opening browser 'gc' to base url 'http://www.baidu.com'.
20190128 17:55:00.229 : INFO : Typing text 'python' into text field 'id=kw'.
20190128 17:55:00.509 : INFO : Clicking element 'id=su'.
20190128 17:55:03.632 : INFO : Slept 3 seconds
20190128 17:55:03.679 : INFO : Current page contains text '英文结果'.
Ending test:   Project.tesAll.Page Should
```

3.13 项目执行顺序

在 Robot Framework 中运行测试项目时是有先后顺序的。首先，从测试套件（文件夹）级别的 Suite Setup 开始执行，如图 3.47 所示。

图 3.47　测试套件（文件夹）Suite Setup 界面

其次，从测试套件（文件）级别的 Suite Setup 开始执行，可以在 testAll 测试套件下的"Suite Setup"选项中自定义一些信息来标识，如图 3.48 所示。

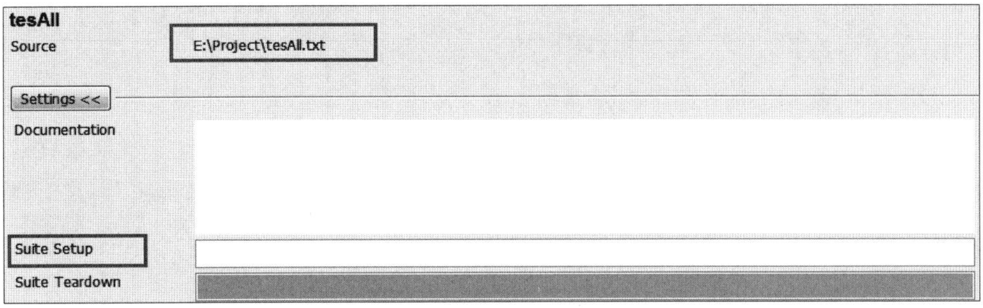

图 3.48　测试套件（文件）Suite Setup 界面

再次，从测试用例级别的 Suite Setup 开始执行，如图 3.49 所示。

图 3.49　测试用例 Setup 界面

最后，进入测试用例开始执行，如图 3.50 所示。

testCase			
1	#打开浏览器		
2	Open Browser	http://www.baidu.com	gc
3	#通过id定位百度文本框		
4	Input text	id=kw	python
5	#单击百度一下按钮		
6	Click Element	id=su	
7	Sleep	3	
8	#断言页面应该包含文本		
9	Page Should Contain	英文结果	
10	#关闭浏览器		
11	Close Browser		

图 3.50　测试用例界面

按 F8 键，查看 RIDE 日志的输出结果：

```
Starting test: Project.tesAll.testCase
20190128 18:26:53.274 : INFO : Opening browser 'gc' to base url 'http://www.baidu.com'.
20190128 18:27:10.522 : INFO : Typing text 'python' into text field 'id=kw'.
20190128 18:27:10.751 : INFO : Clicking element 'id=su'.
20190128 18:27:13.850 : INFO : Slept 3 seconds
20190128 18:27:13.901 : INFO : Current page contains text '英文结果'.
Ending test:   Project.tesAll.testCase
```

从 RIDE 日志结果中不难发现，测试项目在开始运行（Starting test）时，按照 Project（测试项目）→ tesAll（测试套件）→ testCase（测试用例）的顺序进行测试；结束运行（Ending test）时，同样也是按照 Project（测试项目）→ tesAll（测试套件）→ testCase（测试用例）的顺序进行测试。

常见问题整理

在使用 Robot Framework 执行自动化测试的过程中，笔者总结了两类比较常见的问题。其一，有些情况下新增关键字颜色未改变，这时可以通过重新启动 RIDE 来解决。其二，当运行测试脚本时，可能会出现 RIDE 控制台没有日志输出的情况，如图 3.51 所示。

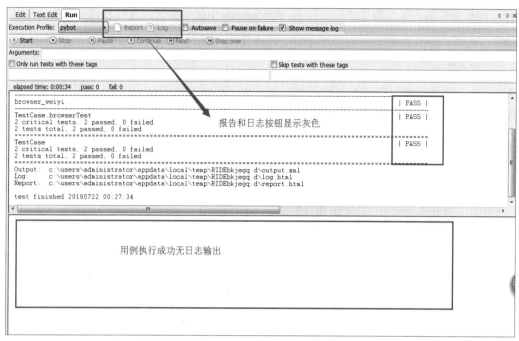

图 3.51　控制台无日志输出界面

用例执行成功，无日志输出且报告和日志按钮显示灰色，原因是开启的 chromedriver.exe 进程没有正常关闭。此时，查看任务管理器中的进程，确认是否有 chromedriver.exe 进程，发现后手动结束进程，即可解决。

也可以通过在 cmd 命令提示符界面中输入结束进程的指令 taskkill /f /im chromedriver.exe 来结束进程，如图 3.52 所示。

图 3.52　chromedriver.exe 进程界面

 ## 自定义关键字

自定义关键字是 Robot Framework 的特性之一，可以理解为函数的封装。自定义关键字的好处是增加了测试案例的重用性和可读性。Robot Framework 中提供的资源文件和自定义用户关键字可以帮助用户实现自定义关键字。

1．创建资源文件

在 Project 测试套件（文件夹）中右击，在弹出的快捷菜单中选择"New Resource File"选项，在弹出的界面中自定义一个名为"Mykeywords"的资源文件，单击"OK"按钮，如图 3.53 所示。

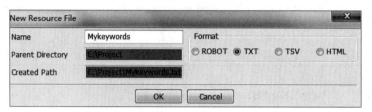

图 3.53　自定义资源文件界面

注意：资源文件名的扩展名可以是 ROBOT、TXT、TSV 和 HTML 等格式，为了管理方便，这里统一使用 .txt 扩展名。

2．创建用户关键字

一个资源文件下可以新增多条用户关键字。右击"Mykeywords.txt"资源文件，在弹出的快捷菜单中选择"New User Keyword"选项，在弹出的界面中自定义一个名为"打开浏览器"的用户关键字，单击"OK"按钮，如图 3.54 所示。

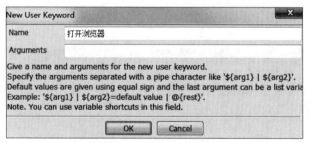

图 3.54　自定义用户关键字界面

3．导入 Selenium2Library 库到资源文件中

单击"Mykeywords.txt"资源文件，然后单击"Library"按钮，导入 Selenium2Library 库，方便后面的测试案例引用该库。

4．新增测试案例

在自定义用户关键字（打开浏览器）中填入测试案例，如图 3.55 所示。

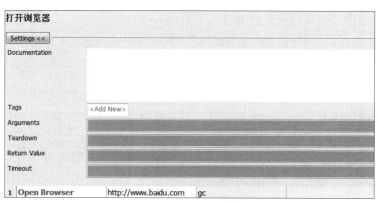

图 3.55　新增测试案例界面

5. 测试用例引用"Mykeywords.txt"资源文件

将"Mykeywords.txt"资源文件导入 testProject 测试套件下的 Resource 中（资源文件导入成功后，资源文件名显示为蓝色），在该测试套件下新增一条测试用例并命名为"test_Search_keys"，将"打开浏览器"用户关键字填入其中，如图 3.56 所示。

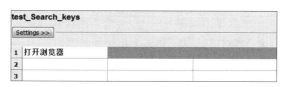

图 3.56　测试用例界面

注意：如果自定义关键字没有显示蓝色，需要将 Selenium2Library 库导入 testProject 测试套件下的 Library 中。

6. 运行 RIDE 查看日志

测试案例运行成功，创建自定义关键字成功。

按 F8 键，查看 RIDE 日志的输出结果：

```
Starting test: Project.testProject.test_Search_keys
20190129 11:18:54.950 : INFO : Opening browser 'gc' to base url 'http://www.baidu.com'.
Ending test:   Project.testProject.test_Search_keys
```

　参数化关键字

在编写了自定义关键字后，可以熟练地调用这个关键字。但是如果关键字的内容有变化，就需要不停地修改这个关键字,这样维护起来就变得很麻烦。

继续对用户关键字"打开浏览器"进行优化。单击创建好的自定义用户关键字(打开浏览器),在"Arguments"选项中定义一个参数变量"${url}",同时在测试案例中引用"${url}"参数变量,如图 3.57 所示。

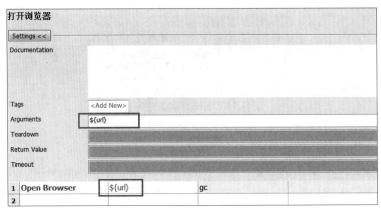

图 3.57　参数化设置界面

如果对多个值进行参数替换,在"Arguments"选项中可以使用"|"进行分隔,如"${name}|${age}|${address}"。同样地,测试用例也需要做出相应的修改,修改后如图 3.58 所示。

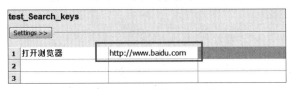

图 3.58　测试用例引用参数界面

下面编写一个登录功能的测试用例。将自定义关键字、关键字参数化等整合到测试用例中,整合后的效果如图 3.59 所示。

图 3.59　测试用例整合界面

在"Mykeywords.txt"资源文件中定义了 5 组用户关键字,分别是打开浏览器、输入搜索内容、单击搜索按钮、验证搜索结果和关闭浏览器。每一组用户关键字下面分别填入对应的测试案例(操作界面的动作),并把测试案例中需要参数化的值进行参数替换。最后将"Mykeywords.txt"导入 testProject 测试套件中,方便 test_Search_keys 测试用例层可以直接引用。

3.17 关键字驱动测试

图 3.59 的测试用例看上去比以前编写的用例可读性提高了很多,但是从用例编写效率来说还存在问题。例如,赘述、冗余的代码太多。试想一下,如果登录测试需要测试 100 个账户,那么是不是要写 100 次这种重复的代码?

关键字驱动测试,实际上是将作用不同的关键字,但是在测试逻辑中又可以通用的部分进行分层封装,使封装后的关键字可以满足各种测试场景及数据的需要,并以最简单明了的方式呈现在测试用例中。

想要用关键字驱动测试,最重要的就是分层封装,而分层封装需要对测试用例进行分解,找出分层规则。

举个登录的例子:

1. 登录流程

    1.1 操作界面

       1.1.1 操作动作

2. 检查登录是否正确

3. 退出或关闭页面

一个完整的登录用例由以上 3 部分组成,登录流程则由一个或多个操作界面组成,操作界面则由多个操作动作组成,操作动作则由一个或多个关键字组成。

根据上面的分析,可以按照以下方式分层:动作层、界面层、流程层和用例层。分层效果演示界面如图 3.60 所示。

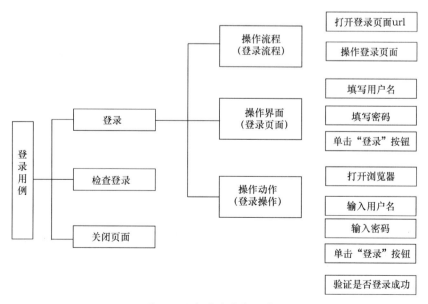

图 3.60 分层效果演示界面

3.18 Settings 界面简介

Settings 界面主要用来配置 TestCase 执行阶段的操作，如数据的初始化、上下文数据还原，以及测试用例结束后数据的销毁、超时设置及模板设置等功能，如图 3.61 所示。

![Settings 界面](settings.png)

图 3.61 Settings 界面

- Documentation：用例说明或关键字说明，主要用在关键字说明中。解释关键字的用途及用法，对各个参数进行说明，无内容限制。
- Setup：执行用例前可进行的一些设置，如数据初始化、上下文数据还原等。
- Teardown：执行用例结束后部分操作，如数据销毁、上下文数据还原等。
- Tags：标签，可用来设置 TestCase 的优先级和标记用例。在用例执行界面（Run）选中 "Only run tests with these tags" 复选框并指定标签名，则运行用例时只会执行有该标签的用例。若未指定标签，则运行所有用例。标签可以指定多个。若未指定标签又想单独运行某个用例，则可以选中用例前的复选框。
- Timeout：超时设置，表示此条用例的最大执行时间。超过此时间，则为失败。
- Template：模板，主要用作数据驱动测试。

3.19 项目分层设计与开发实战

本节使用 Robot Framework 进行自动化测试架构的设计与开发。通过项目分层设计，把自动化测试过程中涉及的操作动作、操作界面、操作流程和测试数据等组织起来，形成一套完备的项目分层架构并坚守。

3.19.1 构建操作动作关键字

在 Project 项目下新建"关键字 - 定义操作动作 .txt"资源文件，并把登录过程中的每个操作动作分别编写成对应的关键字，方便调用，如图 3.62 所示。

图 3.62　关键字 - 定义操作动作界面

在关键字中填入对应操作动作的测试代码。这里以"登录：打开测试页面"关键字为例进行设置，如图 3.63 所示。

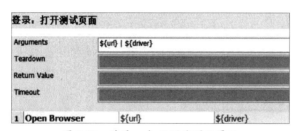

图 3.63　登录：打开测试页面界面

3.19.2 构建操作界面关键字

将操作动作中的关键字按照界面操作顺序进行组合，如图 3.64 所示。

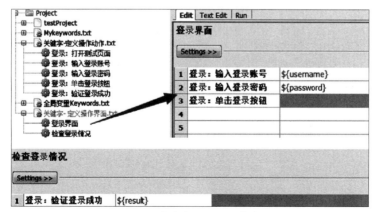

图 3.64　关键字 - 定义操作界面

在 Project 测试项目下新建"关键字 - 定义操作界面 .txt"资源文件，并将 Selenium2Library 库和"关键字 - 定义操作动作 .txt"资源文件导入该资源文件下。在该资源文件下新建"登录界面"和"检查登录情况"关键字，并填入操作的动作。

3.19.3 构建操作流程关键字

将界面及动作中的关键字按照测试操作流程进行组合。在 Project 测试项目下新建"关键字 - 定义操作流程 .txt"资源文件，并导入前面定义好的资源文件，如图 3.65 所示。

图 3.65　资源导入界面

在"关键字 - 定义操作流程 .txt"资源文件下新建关键字并命名为"登录主流程"，填入如图 3.66 所示的信息。

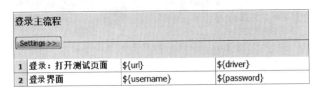

图 3.66　登录主流程界面

在"登录主流程"关键字中引用导入资源文件中的"登录 - 打开测试页面"关键字和"登录界面"关键字，并分别传入对应的参数变量。

3.19.4 构建自动化测试用例

首先在 testProject 测试套件中导入所要用到的"关键字 - 定义操作流程 .txt"资源文件，然后在测试套件下新建一条测试用例并命名为"test_Login"，如图 3.67 所示。

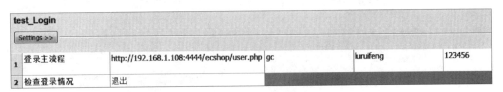

图 3.67　测试用例界面

在测试用例中调用操作流程关键字（登录主流程和检查登录情况），并填入对应的测试数据。

按 F8 键，查看 RIDE 的输出结果：

```
Starting test: Project.testProject.test_Login
2019012916:49:27.542:INFO:Openingbrowser'gc'tobaseurl 'http://192.168.1.108:4444/
ecshop/user.php'.
20190129 16:49:30.660 :   INFO : Typing text 'luruifeng' into text field 'name=username'.
20190129 16:49:30.812 :   INFO : Typing text '123456' into text field 'name=password'.
20190129 16:49:30.951 :   INFO : Clicking element 'name=submit'.
20190129 16:49:34.187 :   INFO : Slept 3 seconds
20190129 16:49:34.486 :   INFO : ${res_text} = 退出
Ending test:    Project.testProject.test_Login
```

需要特殊说明的是，鼠标指针移至关键字同时按 Ctrl 键可以查看关键字说明，了解参数，如图 3.68 所示。

图 3.68　关键字说明界面

3.20 连接 MySQL 数据库实战

DatabaseLibrary 库在 Robot Framework 中用来进行数据库连接及数据查询（本节以 MySQL 为例），通过查询数据返回的结果来对测试用例预期结果进行判断。本节进行 DatabaseLibrary 库的实战。

3.20.1　安装与导入 DatabaseLibrary 库

DatabaseLibray 库的安装分为在线安装和离线安装两种方式，本小节重点介绍在线安装。关于离线安装，只需要在网上下载并解压 DatabaseLibrary 安装包，进入 setup.py 所在文件执行 python setup.py 命令即可。

1. 安装 DatabaseLibrary 库

打开 cmd 命令提示符界面，输入"pip install robotframework-DatabaseLibrary"进行在线安装，示例如下：

```
C:\Users\Administrator>pip install robotframework-databaselibrary
Collecting robotframework-databaselibrary
```

```
Downloading https://files.pythonhosted.org/packages/46/75/969176c4499f435ebd15
45e2d8c5cac797005e6309860db316eda5cb5/robotframework-databaselibrary-1.2.tar.gz
Installing collected packages: robotframework-databaselibrary
  Running setup.py install for robotframework-databaselibrary ... done
Successfully installed robotframework-databaselibrary-1.2
```

2. 导入 DatabaseLibrary 库

在 Project 测试项目下新建 DBTest 测试套件，在该测试套件中导入 DatabaseLibrary 库（区分大小写），如图 3.69 所示。

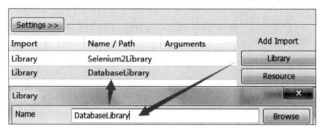

图 3.69　DatabaseLibrary 库导入界面

3.20.2　DatabaseLibrary 库中常用关键字

DatabaseLibrary 库中常用关键字有很多，如连接相关、查询相关和操作相关。这里笔者列出一些较为常用的关键字并进行说明。

（1）连接数据库关键字。

① Connect To Database：连接数据库。

② Connect To Database Using Custom Params：使用自定义参数连接数据库。

（2）关闭数据库关键字。

Disconnect From Database：断开与数据库的连接。

（3）查询数据库关键字。

① Check If Exists In Database：验证数据库中存在查询结果。

② Check If Not Exists In Database：验证数据库中不存在查询结果。

③ Query：返回查询语句的结果。

（4）操作数据库关键字

① Table Must Exist：验证表必须存在，存在则通过，反之，则失败。

② Delete All Rows From Table：删除数据库中表的所有行。

③ Execute Sql Script：执行脚本文件。

3.20.3 连接 MySQL 数据库设置（一）

本小节演示 Robot Framework 使用 Pyodbc 驱动访问数据库的操作方法。

1. 安装 Pyodbc

安装 Pyodbc 包，推荐使用 pip 命令在线安装。打开 cmd 命令提示符界面，输入"pip install pyodbc"进行在线安装，示例如下：

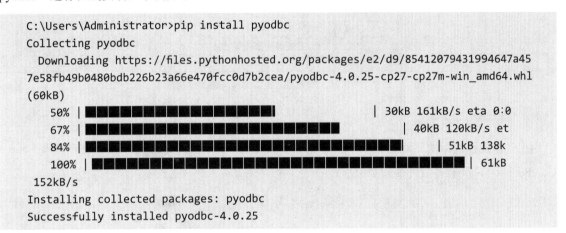

2. 安装 MySQL Connector/ODBC 驱动

MySQL Connector/ODBC 简称 MyODBC。用户可以用 ODBC（Open Database Connectivity，开放数据库互联）数据源连接 MySQL 的服务器。

官方下载地址为 https://dev.mysql.com/downloads/connector/odbc，根据计算机的操作系统版本选择对应的安装包。本小节使用 mysql-connector-odbc-5.3.9-winx64.msi 版本来安装，打开安装界面，单击"Next"按钮，一步步地完成安装即可，如图 3.70 所示。

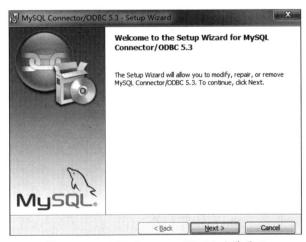

图 3.70　MySQL Connector/ODBC 安装界面

3. 设置 MySQL Unicode Driver

在"控制面板"→"管理工具"中，找到"ODBC 数据源"选项并打开数据源，在"用户 DSN"选项卡中单击"添加"按钮，弹出"创建新数据源"对话框，选择"MySQL ODBC 5.3 Unicode Driver"选项，如图 3.71 所示。

图 3.71　MySQL 数据源添加界面

4. 连接 MySQL 设置

在"创建新数据源"对话框中单击"完成"按钮，弹出 MySQL Connector/ODBC Data Source Configuration 对话框，如图 3.72 所示。

图 3.72　MySQL 驱动设置界面

填入对应参数后，单击"OK"按钮，这样就完成了 Robot Framework 连接 MySQL 数据库的设置过程。

- Data Sourse Name：数据资源名称，这里可以自定义一个名称，如 mysql。
- Description：描述，可以不填。
- TCP/IP Server：数据库地址，127.0.0.1 表示使用本地服务器。
- Port：MySQL 服务的端口号，默认为 3306。
- User：用户名，默认为 root。
- Password：默认为空，如果计算机之前安装过 MySQL，则输入对应的 MySQL 密码。

- Database：选择对应的 MySQL 数据库。
- Test：测试连接是否通过。

3.20.4 连接 MySQL 数据库设置（二）

除可以使用 MySQL 驱动实现 Robot Framework 操作数据库外，还可以使用 pymysql 库来访问数据库。直接安装 pymysql 库，打开 cmd 命令提示符界面，输入 "pip install pymysql" 进行在线安装，示例如下：

```
C:\Users\Administrator>pip install pymysql
Collecting pymysql
  Downloading https://files.pythonhosted.org/packages/ed/39/15045ae46f2a123019aa
968dfcba0396c161c20f855f11dea6796bcaae95/PyMySQL-0.9.3-py2.py3-none-any.whl (47kB)
    42% |█████████████                      | 20kB 89kB/s eta 0:00:01
    64% |████████████████████               | 30kB 120kB/s eta
    85% |██████████████████████████         | 40kB 102k
   100% |███████████████████████████████████| 51kB
116kB/s
Installing collected packages: pymysql
Successfully installed pymysql-0.9.3
```

3.20.5 基于 MySQL 数据库的实战

本小节使用 Robot Framework 连接 MySQL 数据库，并结合案例对 MySQL 数据库进行增、删、改、查等操作。

在操作 MySQL 数据库之前，需要先将 DatabaseLibrary 库导入测试套件中，否则无法引用该库下的关键字。

在 MySQL 数据库中查询指定的数据，如图 3.73 所示。

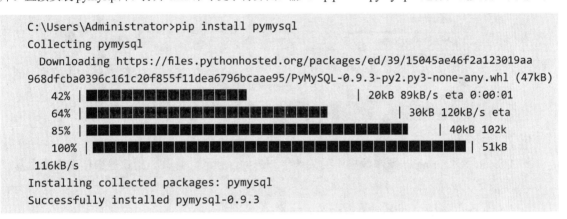

图 3.73　查询数据案例

- Connect To Database Using Custom Params：连接数据库的关键字。引用该关键字需要两个

参数，即 pymysql（连接 MySQL 服务器的库）和数据库连接信息（host=' 数据库 ip', port=' 数据库端口 ', user=' 数据库登录用户名 ', passwd=' 密码 ' 和 db=' 数据库名 (不是表名)'）。
- Query：用于返回查询语句的结果。
- Disconnect From Database：表示与数据库服务器断开连接。

继续向 MySQL 数据库中插入数据，如图 3.74 所示。

	Database_insert插入		
	Settings >>		
1	#连接数据库		
2	Connect To Database Using Custom Params	pymysql	host='127.0.0.1', port=3306, user='root', password='',db='test'
3	#插入数据到数据库		
4	${b}	Execute Sql String	INSERT INTO test.ecs_goods_attr(goods_attr_id,goods_id,attr_id,attr_value) VALUES(335,1,199,'金黄色')
5	Log	${b}	
6	Disconnect From Database		

图 3.74　插入数据案例

- Execute Sql String：操作数据库的关键字。例如，插入、修改和删除等操作都需要用到该关键字来实现。

修改 MySQL 数据库中的数据案例如图 3.75 所示。

	Database_update修改		
	Settings >>		
1	#连接数据库		
2	Connect To Database Using Custom Params	pymysql	host='127.0.0.1', port=3306, user='root', password='',db='test'
3	Execute Sql String	update test.ecs_goods_attr set attr_value = '黑色' where goods_attr_id = 335	
4	${aaa}	Query	SELECT attr_value FROM test.ecs_goods_attr WHERE goods_attr_id=336;
5	Log	${aaa}	
6	Disconnect From Database		

图 3.75　修改数据案例

删除 MySQL 数据库中的指定数据案例如图 3.76 所示。

	Database_delete删除		
	Settings >>		
1	#连接数据库		
2	Connect To Database Using Custom Params	pymysql	host='127.0.0.1', port=3306, user='root', password='',db='test'
3	Execute Sql String	delete from test.ecs_goods_attr where goods_attr_id='335'	
4	${a}	Query	select * from test.ecs_goods_attr where goods_attr_id='335'
5	Should Be Empty	${a}	
6	Disconnect From Database		

图 3.76　删除数据案例

上述案例中，使用 Execute Sql String 关键字删除 test 数据库中 ecs_goods_attr 表下的 goods_attr_id='335' 的商品信息，使用 Query 关键字查询删除后的 goods_attr_id='335' 的商品返回结果，使用 Should Be Empty 关键字断言返回的结果是否为空。

3.21 Jenkins+Robot Framework 持续集成

在实际自动化测试过程中，可以借助 Jenkins 持续平台来自动构建自动化测试任务。本节进行 Jenkins+Robot Framework 持续集成环境的搭建。

3.21.1 安装 robot.hpi 插件

首先确保计算机上已经安装了 Jenkins 环境，关于 Jenkins 的安装可以参考 2.15.8 小节。

robot.hpi 插件用于生成 Robot Framework 自动化测试报告。插件的下载地址为 http://mirrors.jenkins-ci.org/plugins/robot，建议选择 1.3.2 版本。下载完成后，进入 Jenkins 主界面，选择"系统管理"→"管理插件"→"高级"选项，在页面中找到"上传插件"，单击"选择文件"按钮，找到扩展名为 .hpi 的插件，即可将 .hpi 插件上传到 Jenkins 平台并安装，如图 3.77 所示。重启 Jenkins，安装的插件就可以使用了。

图 3.77　robot.hpi 插件上传界面

3.21.2 构建自由风格的任务

在 Jenkins 主界面，新建一个自由风格的任务，然后切换到"构建"选项，在"Execute Windows batch command"界面中的"命令"文本框中输入命令行构建命令，如图 3.78 所示。

如图 3.78 所示，首先切换到 Robot Framework 脚本所在目录（E:\Project），然后通过 pybot 命令执行 DBTest.txt 脚本。

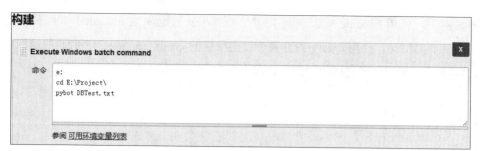

图 3.78　构建设置界面

3.21.3　构建后操作界面设置

进入 Jenkins 设置界面，选择"构建"选项卡，在"构建后操作"界面指定一个 Robot Framework 测试报告输出目录，然后单击"保存"按钮，如图 3.79 所示。

图 3.79　构建后操作设置界面

注意：Robot Framework 输出目录路径要与执行脚本（DBTest.txt）文件所在目录路径保持一致。

3.21.4　查看 Robot Framework 测试报告

在 Jenkins 主界面，选择"立即构建"选项，查看输出结果，如图 3.80 所示。

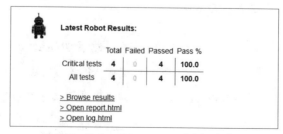

图 3.80　Robot Framework 测试报告界面

第4章

接口测试基础

目前大部分公司的开发模式是前后端分离模式。当后端开发完毕后，接口便已提测，这时前端界面一般还没有进入开发阶段。此时进行接口测试，能提前发现很多系统底层存在的问题，确保前端提测前尽早解决后端存在的问题。本章进行接口测试工作前的接口测试基础理论学习，方便后续章节接口自动化测试任务的实施。

接口测试的定义

接口测试是测试系统组件间接口的一种测试，主要用于检测外部系统与系统之间，以及内部各个子系统之间的交互点。接口测试的重点是检查数据的交换、传递和控制管理过程，以及系统间的相互逻辑依赖关系等。

对计算机而言，接口一般分为两种，一种是程序内部的接口，另一种是系统对外的接口。程序内部的接口是指模块和模块、方法和方法之间的调用。例如，下订单，首先需要登录，然后调用下订单接口，两个模块有交互、相互之间的调用，这就属于内部接口。

系统对外的接口一般是指使用其他公司服务器提供的资源。例如，查询天气预报信息，气象局不会把数据共享出来，但可以提供一个对外的接口，可以引用提供的接口获取天气预报信息，从而达到数据共享的目的。

接口测试的目的

随着互联网产品越来越复杂，从安全性角度考虑，单纯的 UI 层面的功能测试无法覆盖所有参数，存在安全隐患（如某些参数可能会被拦截、篡改）；同时，依靠前端测试会导致测试效率大大降低，现在大部分公司推崇测试前移的思想，就是希望在开发阶段测试能够尽早地介入，而接口测试就是最好的介入方式。在接口测试过程中，开发人员只需要定义好前后端接口，这时测试人员就可以根据接口测试文档、工具或编码提早进行接口测试了。在笔者看来，接口测试的意义有以下几个好处。

（1）检查 UI 界面无法发现的 BUG。

（2）检查系统的安全性和稳定性。

（3）相对 UI 自动化测试，接口测试比较稳定，容易实现自动化持续集成。

（4）尽早发现系统的底层缺陷，降低修复成本。

（5）缩短测试周期，支持后端快速发版的需求。

（6）支持前端的随意变化，后端不用变。

4.3 接口测试原理

接口测试原理可以理解为 HTTP 请求的流程，是指通过测试程序或测试工具模拟客户端向服务器发送请求报文，服务器接收请求报文后对相应的报文做出处理，然后把应答报文发送给客户端这一过程，如图 4.1 所示。

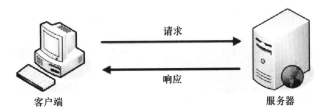

图 4.1 HTTP 请求的流程

4.4 接口测试流程

接口测试本质上属于黑盒测试的范畴，所以接口测试流程和黑盒测试流程大致相同，也需要进行接口需求分析、接口测试用例编写、接口测试工具选择和构造请求并执行测试等工作。下面对各个阶段的任务进行说明。

1. 接口需求分析

接口测试在开发阶段进行，一般需要根据开发提供的接口文档对接口文档内容进行充分研读。接口文档一般包括接口说明、请求地址、请求方式（GET、POST 最为常见）、请求参数、参数类型及参数的约束、正常的响应报文和异常的响应报文等信息。

2. 接口测试用例编写

编写接口测试用例时一般会从以下几个方面来思考：等价类、边界值分析测试，参数组合验证，接口安全验证和业务逻辑验证等。

（1）等价类、边界值分析测试：检查传入的参数合法或不合法、是否为空、参数是否可以包含特殊字符、输入数值范围及数值大小等。例如，查询天气（http://wthrcdn.etouch.cn/weather_mini?city=上海），city 是城市编码，如深圳、上海等。这种接口就要考虑对 city 参数进行等价类、边界值分析测试。

（2）参数组合验证：有时传入的参数分为必传参数和非必传参数，这时要考虑多组排列组合的多种测试情况。例如，快递 100 查询接口 https://www.kuaidi100.com/query?type=yuantong&postid=123456 中，字段 type 表示快递的拼音，字段 postid 表示快递单号。这两个字段都是必传参数，这时就要考虑多组参数组合验证。

（3）接口安全验证：一般为绕过验证，典型的例子就是抓包、修改订单价格。首先，查看后端是否校验；其次，检查参数是否加密，一般是在登录接口检查。例如，如果用户名和密码没有加密，则很容易被别人获取到信息，加密规则很容易被破解等。

（4）业务逻辑验证：存在上下游依赖关系。例如，没有登录就查询订单信息是不可能的，接口之间相互依赖。

3. 接口测试工具选择

现在大部分项目都是基于 HTTP 的接口，所以进行接口测试时主要是通过工具或代码模拟 HTTP 请求的发送与接收。接口测试工具有很多，如 Postman、Jmeter、SoupUI、LoadRunner、Robot Framework+RequestsLibrary、Java+HttpClient 和 Python+Requests 等。接口测试工具推荐使用 Postman 和 Jmeter，两者在安装、使用上更为简单便捷；如果使用编写脚本的方式进行接口测试，推荐采用 Python+Requests 组合。

4. 构造请求并执行测试

根据接口测试用例设计执行测试，检查接口返回的数据是否达到预期。

4.5 接口测试用例设计

接口测试用例设计和黑盒测试用例设计大同小异。接口测试用例的组成由用例名称、前置条件、请求地址、请求 header、请求参数（对应测试数据，即参数）和响应检查点（对应预期结果）等几部分构成。图 4.2 所示为商品查询功能接口测试用例设计界面。

用例名称	前置条件	请求地址	请求header	请求参数	响应检查点
获取所有商品的sku列表成功	无	GET /common/skushopList	Content-Type=application/json	无	code : 200 message : "success" result : *所有商品sku信息列表(略)*
获取goodsId=1的商品sku信息成功	无	GET /common/skushopList	Content-Type=application/json	goodsId=1	code : 200 message : "success" result : *goodsId=1的商品sku信息(略)*
获取goodsId=55421322的商品sku信息失败(超过int最大取值范围)	无	GET /common/skushopList	Content-Type=application/json	goodsId=55421322	code : 400 message : "商品ID不正确"
获取goodsId不存在的商品失败	无	GET /common/skushopList	Content-Type=application/json	goodsId=9999	code : 201 message : "商品ID不存在"
goodsId类型不正确	无	GET /common/skushopList	Content-Type=application/json	goodsId="1"	code : 400 "message" : "商品ID参数类型不正确"

图 4.2　商品查询功能接口测试用例设计界面

图4.2中，针对商品查询接口功能，使用等价类、边界值分析测试方法设计了5条接口测试用例，分别是获取所有商品、查询指定商品、查询商品编号超过最大取值范围的商品、查询不存在的商品和输入商品参数不正确（不合法参数校验）。

4.6 HTTP 基础

HTTP（HyperText Transfer Protocol，超文本传输协议）是用于从 WWW 服务器传输超文本到本地浏览器的传送协议，可以使浏览器更加高效，减少网络传输。HTTP 是一个应用层协议，由请求和响应构成，是一个标准的客户端服务器模型，也是一个无状态的协议。

4.6.1 HTTP 请求报文

客户端向服务端发送请求报文（GET 或 POST 请求）时，请求报文由 4 部分组成，分别是请求行、请求头信息、空行和请求体。

GET 请求一般是向服务器获取指定资源，使用 Fiddler 抓包进一步分析 GET 请求信息，如图 4.3 所示。

```
Statistics | Inspectors | AutoResponder | Composer | Fiddler Orchestra Beta | FiddlerScript | Log | Filters | Timeline
Headers | TextView | SyntaxView | WebForms | HexView | Auth | Cookies | Raw | JSON | XML
GET http://192.168.0.162:4444/zentao/www/my/ HTTP/1.1
Host: 192.168.0.162:4444
Connection: keep-alive
Upgrade-Insecure-Requests: 1
User-Agent: Mozilla/5.0 (Windows NT 6.1; Win64; x64) AppleWebKit/537.36 (KHTML, like Gecko) Chrome/70.0.3538.110 Safari/537.36
Accept: text/html,application/xhtml+xml,application/xml;q=0.9,image/webp,image/apng,*/*;q=0.8
Referer: http://192.168.0.162:4444/zentao/www/user-login-L3plbnRhby93d3cv.html
Accept-Encoding: gzip, deflate
Accept-Language: zh-CN,zh;q=0.9
Cookie: lang=zh-cn; theme=default; windowWidth=1280; windowHeight=631; sid=7jcao0t8apefn43aaprqphfgt1
```

图 4.3　GET 请求案例

（1）请求行：包含请求方法 GET、请求地址 http://192.168.0.162:4444（访问本地服务器地址）和 HTTP 版本号，如 HTTP/1.1。

（2）请求头信息：包含 User-Agent、Accept、Accept-Encoding 和 Accept-Lanuage 等头字段信息。

（3）空行：请求头信息后面的空行是必须存在的，没有数据。

（4）请求体：一般没有数据。请求数据都会在地址栏内显示，本案例中请求数据为空。

POST 请求一般是向服务器发送数据。图 4.4 所示为 POST 请求登录的案例。

```
POST http://192.168.0.162:4444/zentao/www/user-login.html HTTP/1.1
Host: 192.168.0.162:4444
Connection: keep-alive
Content-Length: 120
Cache-Control: max-age=0
Origin: http://192.168.0.162:4444
Upgrade-Insecure-Requests: 1
Content-Type: application/x-www-form-urlencoded
User-Agent: Mozilla/5.0 (Windows NT 6.1; Win64; x64) AppleWebKit/537.36 (KHTML, like Gecko) Chrome/70.0.3538.110 Safari/537.36
Accept: text/html,application/xhtml+xml,application/xml;q=0.9,image/webp,image/apng,*/*;q=0.8
Referer: http://192.168.0.162:4444/zentao/www/user-login.html
Accept-Encoding: gzip, deflate
Accept-Language: zh-CN,zh;q=0.9
Cookie: lang=zh-cn; theme=default; lastProduct=3; productStoryOrder=id_desc; lastBranch=3-0; qaBugOrder=id_desc; windowHeight=631;
account=admin&password=123456&referer=http%3A%2F%2F192.168.0.162%3A4444%2Fzentao%2Fwww%2Fbug-browse-3-0-byModule-15.html
```

图 4.4　POST 请求案例

（1）请求行：包含请求方式 POST、请求地址及 HTTP 版本号。

（2）请求头信息。包括以下内容。

① Accept-Encoding：指定浏览器可以支持的服务器并返回内容压缩编码类型，如 gzip。

② Accept：指定客户端接收哪些类型的信息，如 Accept:text/html，表明客户端希望接收 HTML 文本。

③ Accept-Language：用于指定一种自然语言，如 Accept-Language:zh-cn。如果请求消息中没有设置这个报头域，则服务器假定客户端对各种语言都可以接收。

④ Content-Type：请求与实体对应的 MIME 媒体信息，一般以 POST 方式提交数据常见的几种 Content-Type 类型。

- application/x-www-form-urlencoded：原生 form 表单提交方式，数据会自动被编码为键值对。
- multipart/form-data：支持在表单中进行文件上传。
- application/json：以 JSON 字符串提交数据。
- text/xml：以 XML 数据格式提交数据。

（3）空行：请求头信息后面的空行是必须存在的，没有数据。

（4）请求体：包含请求数据，如账号、密码都是加密后显示的。

4.6.2　HTTP 响应报文

服务端接收客户端请求后，会返回客户端响应报文。响应报文由 4 部分组成：响应状态行、响应头信息、空行和响应体。下面来看一个 HTTP 响应案例，如图 4.5 所示。

```
HTTP/1.1 200 OK
Date: Wed, 19 Dec 2018 07:49:58 GMT
Server: Apache/2.2.21 (Win32) mod_ssl/2.2.21 OpenSSL/1.0.0e PHP/5.3.8
X-Powered-By: PHP/5.3.8
Set-Cookie: lang=zh-cn; expires=Fri, 18-Jan-2019 07:49:58 GMT; path=/zentao/www/
Set-Cookie: theme=default; expires=Fri, 18-Jan-2019 07:49:58 GMT; path=/zentao/www/
Expires: Thu, 19 Nov 1981 08:52:00 GMT
Cache-Control: private
Pragma: no-cache
Content-Length: 82
Keep-Alive: timeout=5, max=100
Connection: Keep-Alive
Content-Type: text/html; Language=UTF-8

<html><head><meta http-equiv='refresh' content='600' /></head><body></body></html>
```

图 4.5　HTTP 响应案例

（1）响应状态行：包含 HTTP 版本号和响应状态码（200 表示请求成功）。

（2）响应头信息：包含响应信息时间、缓存、过期时间和响应实体正文的媒体类型等信息。

① Date：表示消息发送的时间。

② Server：表示服务器用来处理请求的软件信息，如服务器版本和开发语言等。

③ X-Power-By：用于告知网站是用何种语言或框架编写的。

④ Set-Cookie：服务器以 Set-Cookie 方式向浏览器写入 Cookie 信息。

⑤ Expires：指定页面过期的时间，目的是缩短再次访问曾访问过的页面的响应时间。

⑥ no-cache：用于指示请求或响应消息不能缓存。

⑦ Content-Length：用于指明实体正文的长度，以字节方式存储的十进制数字来表示。

⑧ Connection：客户端和服务器连接都是 Keep-Alive（持久连接、连接重用）持续有效，当出现对服务器的后继请求时，Keep-Alive 功能避免了建立或重新建立连接。

⑨ Content-Type：指明发送给接收者的实体正文的媒体类型。text/html 表示返回的内容是文本类型，这个文本是 HTML 格式的。

（3）空行：响应头信息后面的空行必须存在。

（4）响应体：HTML 这部分代码是服务器响应客户端的正文。

4.6.3　HTTP 状态码

HTTP 状态码（HTTP Status Code）是表示网页服务器超文本传输协议响应状态的 3 位数字代码。所有状态码的第一个数字代表了响应的 5 种状态之一。

（1）状态码分类。

① 1xx：服务端收到请求，需要进一步处理，客户端请等待。

② 2xx：请求处理成功，常见状态码为 200。

③ 3xx：重定向操作，需要进一步操作完成请求。

④ 4xx：客户端请求错误，请求有语法错误或请求无法实现。

⑤ 5xx：服务器内部错误，不能正确处理请求。

(2)常见状态码。

① 200（OK）：客户端请求成功。

② 301（Moved Permanently）：请求地址永久转移到新的 URL，浏览器会自动定向到新 URL。

③ 302（Found）：请求地址临时转移，与 301 类似。

④ 400（Bad Request）：客户端请求的语法错误，服务器无法理解。

⑤ 401（Unauthorized）：请求要求用户的身份认证，未经授权。

⑥ 403（Internal Server Error）：服务器理解请求客户端的请求，但是拒绝服务。

⑦ 404（Internal Server Error）：请求资源不存在。

⑧ 500（Internal Server Error）：服务器内部错误，无法完成请求。

4.6.4 HTTP 请求方法

HTTP 请求方法有多种，实际项目中比较常用的是 PUT、DELETE、POST 和 GET。这几种请求方法针对服务器资源而言分别对应增加、删除、修改和查询操作。下面介绍这些请求方法的含义。

（1）PUT：从客户端向服务器传送的数据取代指定文档的内容。

（2）DELETE：用于删除服务器指定的资源。

（3）POST：用于向服务器提交数据（如提交表单或上传文件），请求数据一般在请求体中，请求可能会导致新资源的建立或已有资源的修改。

（4）GET：用于向服务器获取资源，返回响应实体的内容。

4.7　Cookie 和 Session

会话（Session）跟踪是 Web 程序中常用的技术，用来跟踪用户的整个会话。常用的会话跟踪技术是 Cookie 与 Session。Cookie 通过在客户端记录信息以确定用户身份，Session 通过在服务器端记录信息以确定用户身份。

4.7.1　Cookie 的工作原理

Cookie 是当客户端浏览器向服务器发送请求信息时，服务器存储在客户端（一般指浏览器）上的一个文本文件，记录用户 ID、密码、浏览过的网页和停留时间等信息，当浏览器再次发送请求给服务器时，服务器通过读取 Cookie 得知客户端的相关信息，就可以做出相应的动作。例如，在页面显示欢迎你的标语，或者让客户端不用输入 ID、密码就直接登录等。

Cookie 可以保持登录信息到用户下次与服务器进行会话，换句话说，下次访问同一网站时，用户会发现不必输入用户名和密码就已登录（当然不排除用户手动删除 Cookie）。而还有一些 Cookie 在用户退出会话时就被删除，这样可以有效保护个人隐私。

Cookie 在生成时就会被指定一个 Expire 值，即 Cookie 的生存周期，在这个周期内 Cookie 有效，超出周期 Cookie 就会被清除。有些页面将 Cookie 的生存周期设置为"0"或负值，在关闭浏览器时自动清除 Cookie，不会记录用户信息，更加安全。

4.7.2　Session 的工作原理

Session 机制是一种服务器端的机制，服务器使用一种类似于散列表的结构（也可能就是使用散列表）来保存信息。

与 Cookie 不同的是，Cookie 保存在客户端浏览器中，而 Session 保存在服务器中。当客户端浏览器访问服务器时，服务器会根据请求信息生成与 Session 相关联的 session_id（session_id 的值是一个既不会重复，又不容易被找到规律以仿造的字符串，保存这个 session_id 的方式可以采用 Cookie）。服务器将 session_id 通过 Cookie 的方式发送给客户端，当客户端再次向服务端发送请求时，会通过 Cookie 将这个 session_id 发送给服务端，这样就可与 Session 关联上，即 Session 是基于 Cookie 的。

第5章
Charles抓包工具实战

Charles 是一款非常实用方便、界面友好的抓包工具。它基于 Java 语言开发，可跨平台支持 Liunx、Windows 和 Mac 操作系统。Charles 抓包相当于一个处在客户端（PC 或移动端）和服务器之间的"过滤器"，可以捕获客户端发送到服务器的所有请求。

5.1 下载与安装 Charles

因为 Charles 是基于 Java 语言开发的，所以在使用 Charles 工具前，计算机需要安装 Java 环境。

访问 Charles 官网（http://www.charlesproxy.com），下载 Charles 工具。Charles 工具是一个收费软件，使用有效期默认 30 天，但在使用时一般每隔 30min 就会自动关闭退出，建议读者如果要长期使用，可以去注册购买正版软件。

5.2 计算机端抓包设置

1. 在计算机端安装证书

（1）打开 Charles，选择"Help"→"SSL proxying"→"Install Charles Root Cerfficate"选项，弹出"证书"对话框，如图 5.1 所示。

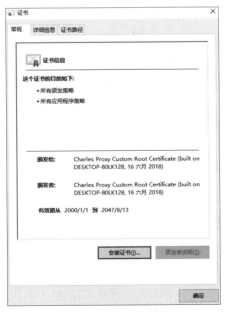

图 5.1　安装证书界面

（2）单击"安装证书"按钮，弹出"证书导入向导"对话框，单击"下一步"按钮；选中"将所有的证书放入下列存储"单选按钮，然后单击"浏览"按钮，在弹出的"选择证书存储"对话框中选择"受信任的根证书颁发机构"选项，如图 5.2 所示。

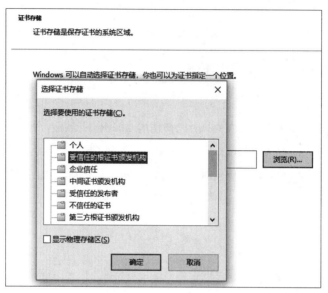

图 5.2　证书导入界面

（3）单击图 5.2 中的"确定"按钮，再单击"下一步"按钮，最后单击"完成"按钮，弹出"导入成功"提示框，如图 5.3 所示。

图 5.3　证书导入成功界面

2. 配置 SSL Proxying 代理

打开 Charles，选择"Proxing"→"SSL Proxying Settings"→"SSL Proxying"选项，在"SSL Proxying"选项卡中选中"Enable SSL Proxying"复选框，单击"Add"按钮，弹出"Edit Location"对话框，在"Host:"文本框中输入"*"（*表示抓取任意站点），在"Port:"文本框中输入"443"（443是默认端口号），然后单击"OK"按钮，如图 5.4 所示。

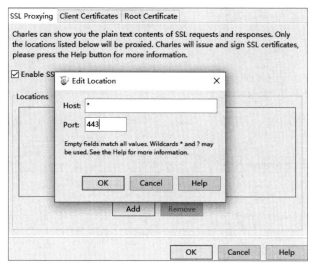

图 5.4　Add Locaitons 界面设置

3. 抓取 HTTPS 请求

Charles 可以抓取 HTTP 和 HTTPS 的请求，但不支持抓取 Scoket 请求。本节演示使用 Charles 工具抓取人人网登录 POST 请求案例，如图 5.5 所示。

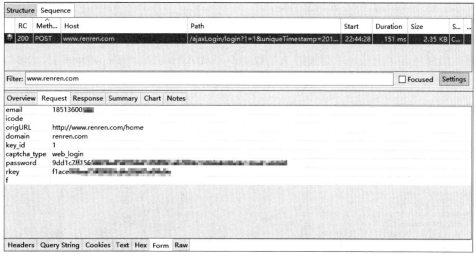

图 5.5　人人网登录 POST 请求案例

如图 5.5 所示，Request 代码是请求的参数。选择"Response"选项，查看服务器响应报文，如图 5.6 所示。

通过响应报文，可以看到服务器返回的是 JSON 数据。选择"Request"→"Headers"选项，查看请求头信息，如图 5.7 所示。

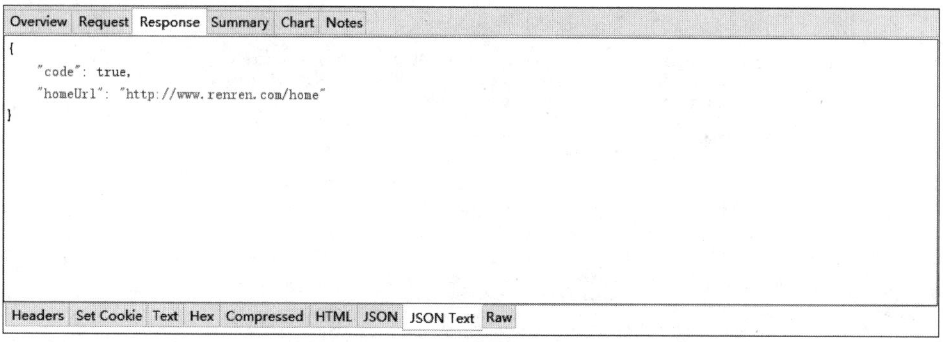

图 5.6 响应报文界面

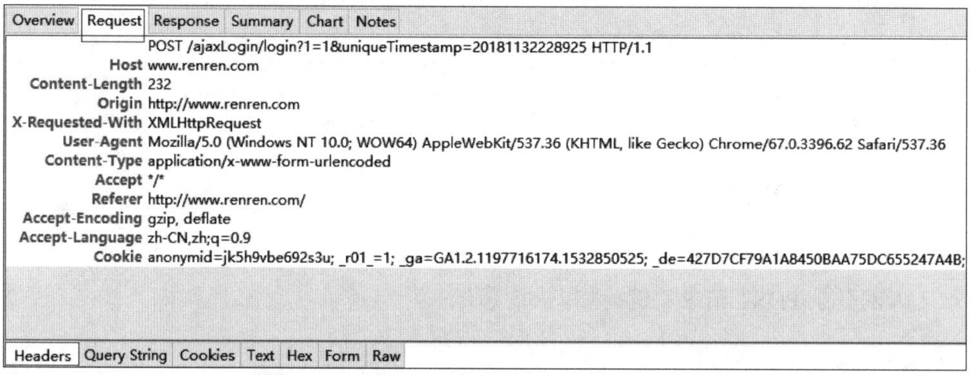

图 5.7 请求头信息

请求头信息包含了一些比较重要的头字段，如 User-Agent、Accept-Encoding、Accept-Language、Content-Type 和 Content-Length 等。关于这部分内容可以参考 4.6.2 小节。

选择"Response"→"Headers"选项，查看响应头信息，如图 5.8 所示。

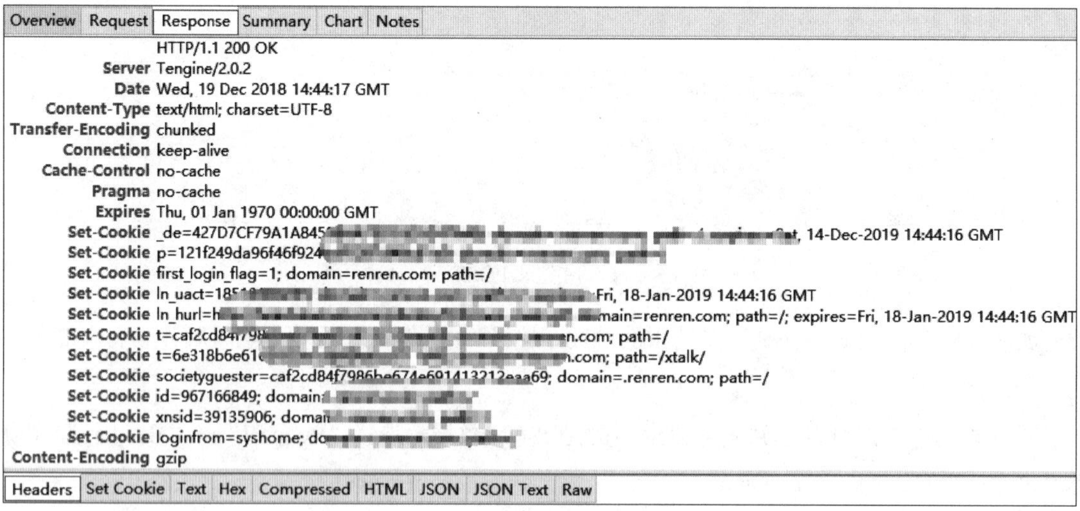

图 5.8 响应头信息

5.3 手机端抓包设置

1. Charles 上的设置

如果抓取手机端上的包，需要将 Charles 代理功能打开。打开 Charles，选择"Proxy"→"Proxy Settings"选项，弹出"Proxy Settings"对话框，在其"Port:"文本框中输入"8888"（默认存在的），并选中"Enable transparent HTTP proxying"复选框即可完成在 Charles 上的设置，如图 5.9 所示。

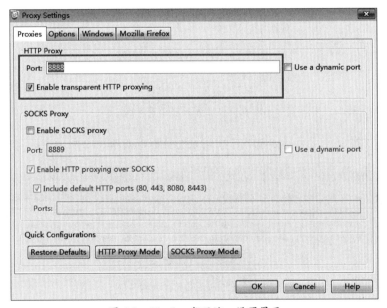

图 5.9　Charles 代理端口设置界面

2. 手机上的设置

（1）获取计算机 IP 地址。打开 Charles，选择"Help"→"Local IP Address"选项，即可在弹出的对话框中看到 IP 地址，如图 5.10 所示。

图 5.10　查看计算机 IP 地址

（2）打开手机端的 WiFi 代理设置，可以看到当前连接的 WiFi 名称，点击最右边的扩展箭头，可以看到当前所连接的 WiFi 信息，包括 IP 地址、子网掩码等。找到最底部"HTTP 代理"选项，将默认关闭切换成手动，在"服务器"处输入计算机的 IP 地址（图 5.10），在"端口"处输入

"8888"，如图 5.11 所示。

图 5.11　配置代理设置界面

（3）设置完代理，在计算机上会弹出手机请求对话框，如图 5.12 所示。

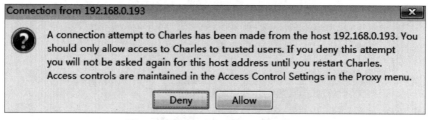

图 5.12　Charles 配置代理确认界面

（4）单击"Allow"按钮，即可完成配置。如果在手机端抓取 HTTPS 请求，需要在 Charles 中选择"Help"→"SSL Proxying"→"Install Charles Root Certificate on a Mobile Device or Remove Browser"选项，在弹出的对话框中单击"确定"按钮，如图 5.13 所示。

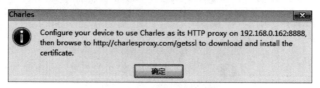

图 5.13　证书安装确认对话框

（5）Charles 证书的下载安装地址为 http://www.charlesproxy.com/getssl，配置好代理后，可以在手机浏览器中输入以上地址来安装 Charles 证书，如图 5.14 所示。

安装完成后即可抓取手机端的 HTTPS 请求。

图 5.14 证书安装界面

 Charles 过滤请求

抓包时 Charles 会捕获很多不必要的请求，这时就需要对网络请求进行过滤，抓取需要的目标请求地址。在 Charles 主界面的中部的 "Filter:" 文本框中输入需要过滤出来的关键字。例如，服务器的地址为 www.renren.com，只需要在 "Filter:" 文本框中输入 "www.renren.com" 即可，这样就可以只保留以 www.renren.com 开头的请求，如图 5.15 所示。

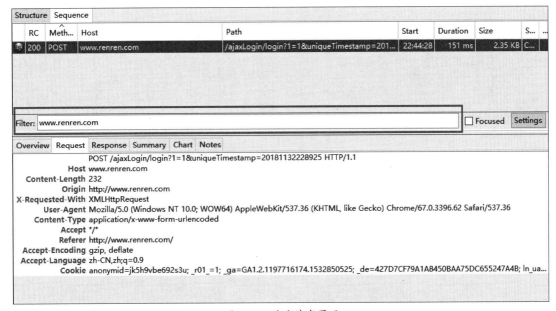

图 5.15 过滤请求界面

5.5 Charles 常见问题

在使用 Charles 抓包时，经常会遇到很多问题，笔者列举了以下一些比较常见的问题。

（1）Charles 下载后无法使用。因为 Charles 由 Java 语言开发，所以需要先安装 Java 环境才可以使用。

（2）使用 Charles 大概 30min 就会关闭一次。因为 Charles 如果没有注册，所以每次打开后就只能用 30min，然后就会自动关闭。这是正常现象，可以注册购买正版。

（3）使用 Charles 后无法上网。如果 Charles 是非正常状态下关闭的，那么 IE 的代理就不会被自动取消，必然会导致这种情况。

（4）使用 Charles 抓取手机 APP，配置正确但却抓不到数据，防火墙是关闭状态。

（5）使用 Charles 抓取手机 APP 的数据，但同时也会抓取到计算机端的数据。选择"Proxy"→"Windows Proxy"选项进行设置，选中表示接收计算机的数据抓包（如果只想抓取 APP 的数据请求，可以不选中此功能）。

第6章

Postman接口测试高级实战

Postman 是一款非常好用、功能强大的网页调试与发送网页 HTTP 请求的 Chrome 插件，适用于不同的操作系统，如 Mac、Windows、Linux 操作系统，同时还支持 Postman 浏览器扩展程序、Postman Chrome 应用程序等。由于 2018 年年初 Chrome 停止对 Chrome 应用程序的支持，所以 Postman 插件可能无法正常使用了。目前 Chrome 应用商店使用的是 Chrome 扩展程序和主题背景。如果想用 Postman，可以使用官方推出的 Postman 开发者版本。

安装 Postman

Postman 的安装非常简单，打开网址 https://www.getpostman.com/apps 下载即可。注意区别操作系统版本，本节基于 Windows 10（x64）、Postman 工具（Version 5.5.0）演示。安装完成后，打开 Postman，界面如图 6.1 所示。

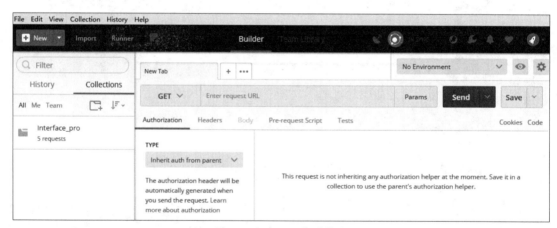

图 6.1　Postman 首页界面

Collections 简介

Collections 可以看作用来存放一个测试项目中所有接口测试用例的根目录文件夹。在 Collections 文件夹中可以创建多个二级子文件夹（文件夹不要使用中文命名），每个二级子文件夹中都可以存放一条或多条接口测试用例，如图 6.2 所示。

第 6 章
Postman 接口测试高级实战

图 6.2　Collections 界面

在 Collections 文件夹中创建 Interface_pro 目录，Interface_pro 可以作为一个接口项目来管理所有接口测试用例。在 Interface_pro 主目录下新增 InterApi 二级目录，InterApi 可以用来表示接口项目中的某一个接口功能，在 InterApi 文件夹下编写多条接口测试用例。

在接口地址栏内添加请求地址为 http://www.baidu.com，请求方式为 GET，单击 "Send" 按钮发送，如图 6.3 所示。

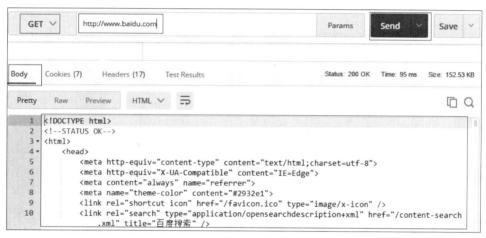

图 6.3　百度响应信息界面

- Body：显示服务器响应的正文信息。
- Status: 200 OK：表示请求处理成功。
- Time: 95 ms：表示响应信息花费的时长为 95ms。
- Size: 152.53 KB：表示响应数据大小为 152.53KB。

单击 "Save" 按钮，将接口保存到 InterApi 文件夹中。

基于 HTTP 接口实战

本节使用 Postman 工具处理接口测试中 POST 请求发送常用的几种数据格式，如 application/

x-www-form-urlencoded、application/json 等，以及结合实际项目演示 Cookie 和 Session 在接口测试过程中的应用。

6.3.1 处理原生 Form

在 Interface_pro 项目下创建 InterApi_POST 子文件夹，用来存放 POST 请求。以禅道（一款开源项目管理软件）为例，禅道登录方式是采用原生 Form 表单方式提交，即 application/x-www-form-urlencoded 数据格式，如图 6.4 所示。

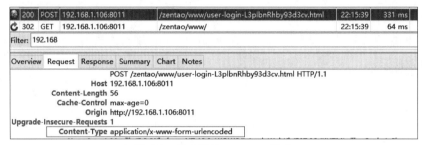

图 6.4　禅道登录 Requests 界面

使用 Charles 抓取禅道登录信息，如图 6.5 所示。

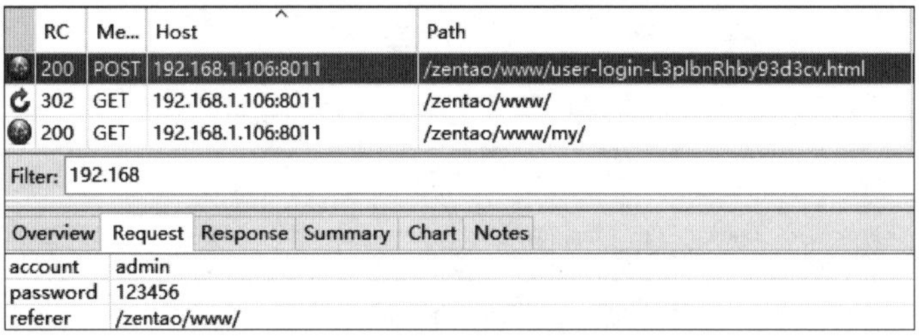

图 6.5　禅道登录地址界面

在 Headers 中新增请求头信息，即 Content-Type=application/x-www-form-urlencoded，如图 6.6 所示。

图 6.6　Postman 请求头信息界面

选择"Body"选项卡中的"x-www-from-urlencoded"选项，将禅道登录请求信息：请求方法、请求地址和请求参数分别填入 Key 和 Value 列表框中，然后单击"Send"按钮，如图 6.7 和图 6.8 所示。

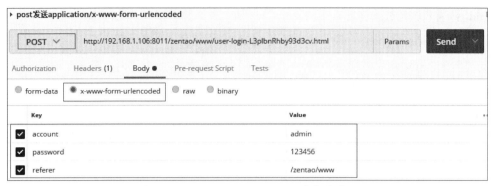

图 6.7　x-www-form-urlencoded 请求体界面

图 6.8　x-www-form-urlencoded 响应体界面

6.3.2　处理 JSON

有些接口数据采用 JSON 数据格式，在向服务器发送 JSON 请求参数时，需要在请求头信息 Headers 中添加 Content-Type=application/json 来声明，如图 6.9 所示。

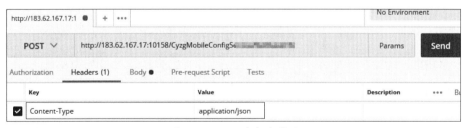

图 6.9　JSON 请求头界面

在"Body"选项卡中填入 JSON 数据，如图 6.10 所示。

图 6.10　JSON 请求体界面

- raw：表示原始数据。
- JSON(application/json)：表示传递的是 JSON 数据格式。

单击"Send"按钮，查看服务器响应信息，如图 6.11 所示。

```
{
    "ErrorInfo": "城市信息获取成功",
    "ResultCode": 10001,
    "ResultInfo": "[{\"province_id\":\"     \",\"province_name\":\"    \",\"cities\":[{\"city_id\"
    :\"430100\",\"city_name\":\"    \",\"is_appoint\":\"0\",\"isNotCheckFace\":\"0\"
    ,\"appoint_Url\":\"\",\"examOverValue\":0,\"mnksOverValue\":60,\"dycsOverValue\":60,\"server_Url\"
    :\"http://node1.m.jtcyzg.cn:10150/Changsha/CyzgMobileService.svc/\",\"exam_Url\":\"\",\"fileLoad_Url\"
    :\"http://jxjy-app-osc-cn-shenzhen.aliyuncs.com/app_setup/JXJY_Phone_v1.1.2.apk\",\"mrServer_Url\"
    :\"http://1        2:10108/CyzgTaxiService.svc/\",\"mrFileLoad_Url\":\"http://        :10286
    /ChapterFlvHandler.ashx/\",\"helpVedio_Url\":\"http://1       9:10026/ChapterFlvHandler.ashx
    /HelpVedio/0.mp4\",\"IsNeedLearnHour\":\"0\",\"jxjyHttpsUrl\":\"https://node1.m.jtcyzg.cn:10151/Test
    /CyzgMobileService.svc/\",\"kqpxHttpsUrl\":\"\",\"jxjyRegisterUrl\":\"\",\"kqpxRegisterUrl\":null
    ,\"isOpenHuXin\":0,\"isExamOnline\":0,\"isEncrypted\":false,\"CustomerServicesInfos\":[{\"name\"
    :\"QQ客服\",\"oicq\":\"    72\",\"serviceTime\":\"09:00-21:00\",\"serviceType\":1,\"isLoginShow\"
    :1},{\"name\":\"QQ客服\",\"oicq\":\"       5\",\"serviceTime\":\"09:00-21:00\",\"serviceType\":1
    ,\"isLoginShow\":0},{\"name\":\"QQ客服\",\"oicq\":\"3321341510\",\"serviceTime\":\"09:00-21:00\"
    ,\"serviceType\":1,\"isLoginShow\":0},{\"name\":\"QQ客服\",\"oicq\":\"3356596426\",\"serviceTime\"
    :\"09:00-21:00\",\"serviceType\":1,\"isLoginShow\":0},{\"name\":\"微信客服\",\"oicq\":\"k     2\"
    ,\"serviceTime\":\"09:00-21:00\",\"serviceType\":2,\"isLoginShow\":1},{\"name\":\"服务热线\",\"oicq\"
    :\"      \",\"serviceTime\":\"09:00-21:00\",\"serviceType\":4,\"isLoginShow\":1},{\"name\"
    :\"微信公众号\",\"oicq\":\"校谈车生活\",\"serviceTime\":\"09:00-21:00\",\"serviceType\":3
    ,\"isLoginShow\":0}]}]}]"
}
```

图 6.11　JOSN 响应数据界面

6.3.3　处理 Cookie

以禅道举例，如果不登录禅道是无法在禅道中创建产品信息的，但可以通过 Cookie 来实现登录过程。获取 Cookie 的整个思路是先登录禅道，登录成功后获取服务器返回给客户端 Cookie，再把 Cookie 作为第二次发送请求时（创建产品接口）的参数。

（1）使用 Charles 抓包获取登录系统服务器返回的 Cookie 信息，如图 6.12 所示。

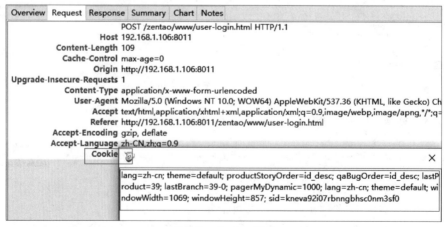

图 6.12　Cookie 界面

（2）在创建产品接口请求头 Headers 中填入服务器返回的 Cookie 信息，再将请求参数信息填入 Body 中，如图 6.13 所示。

图 6.13 填入请求体参数界面

单击"Send"按钮，查看服务器响应信息，如图 6.14 所示。

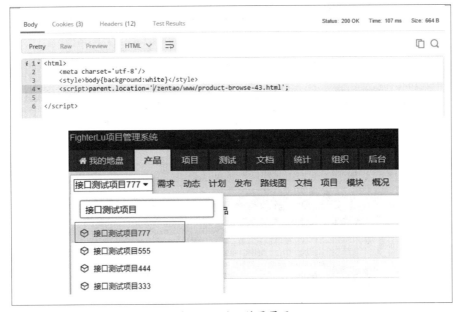

图 6.14 验证结果界面

6.3.4 处理 Session

关于 Session 的介绍可以参考 4.7.2 小节。以博客园为例，在进行接口测试的过程中，不允许直接访问粉丝详情。首先登录博客园，登录成功后才可以查看粉丝详情。博客园的登录是基于Session实现的，获取Session的思路和Cookie大同小异，都是先登录系统，获取登录后服务器返回的Session，然后将Session作为下一个请求的参数。

使用Charles抓取博客园登录请求，获取服务器返回的Session信息，如图6.15所示。

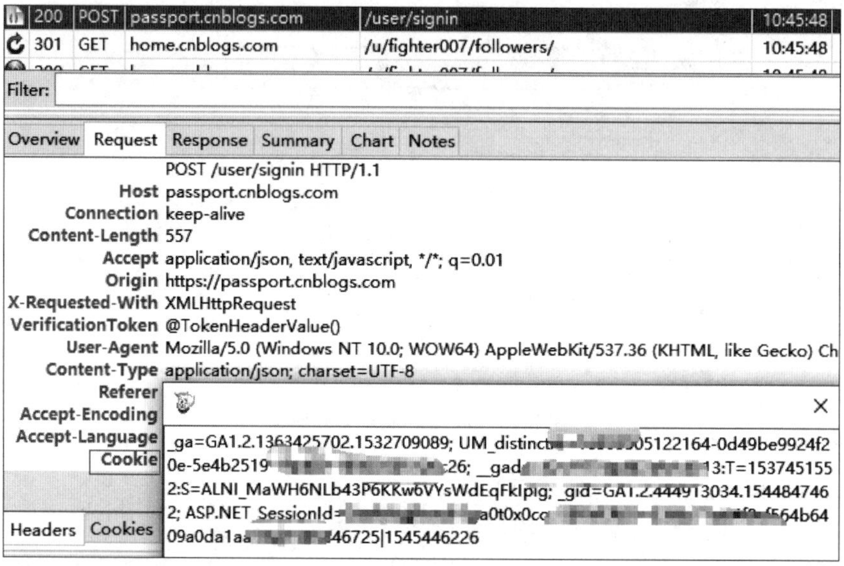

图 6.15 Session 界面

查看博客园粉丝详情接口系统自动带上登录后服务器返回的Cookie，如图6.16所示。

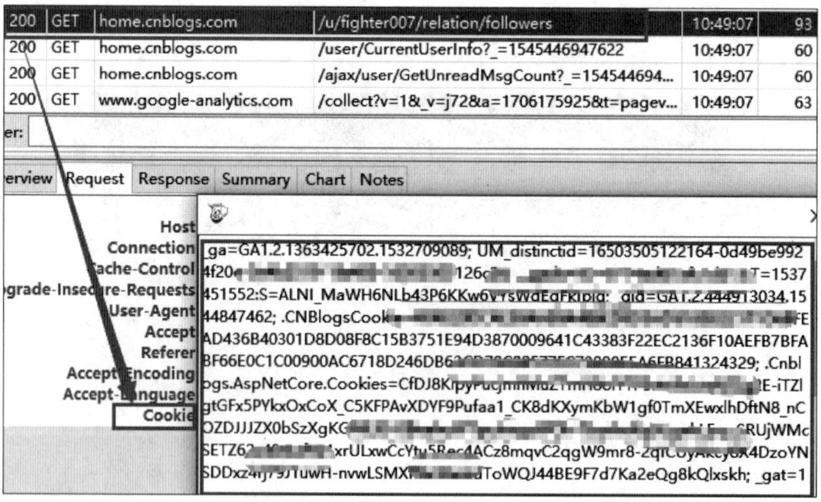

图 6.16 查看粉丝请求的 Cookie 界面

通过Response可以看出博客园粉丝详情是访问成功的，如下所示：

```
<!DOCTYPE html>
<html>
```

```
<head>
    <meta charset="utf-8" />
    <meta name="viewport" content="width=device-width,initial-scale=1,maximum-
scale=1.0,minimum-scale=1.0,user-scalable=no" />
    <title> 我的粉丝 - 我的园子 - 博客园 </title>
    <link href="//common.cnblogs.com/favicon.ico" rel="shortcut icon" />
    <linkrel="stylesheet"href="/css/reset.min.css?v=NNKSv2cA90mrhu98reNDRA-Me_
BeHP9fhzvfTv39pb0" />
    <linkrel="stylesheet"href="/css/home_common.css?v=8h9V8hkCCajAlYCQTOkqgF-
CYDhdI0sknRaUIsm8dI8" />
```

使用 Postman 工具实现查看粉丝详情。先通过 Charles 抓取博客园查看粉丝的接口，如图 6.17 所示。

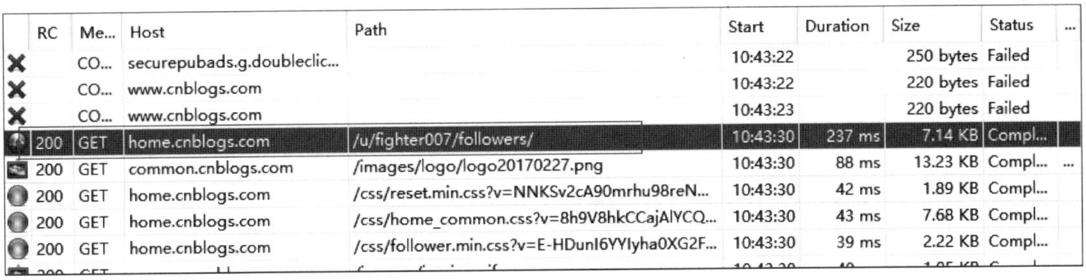
RC | Me... | Host | Path | Start | Duration | Size | Status
CO... securepubads.g.doubleclic... 10:43:22 250 bytes Failed
CO... www.cnblogs.com 10:43:22 220 bytes Failed
CO... www.cnblogs.com 10:43:23 220 bytes Failed
200 GET home.cnblogs.com /u/fighter007/followers/ 10:43:30 237 ms 7.14 KB Compl...
200 GET common.cnblogs.com /images/logo/logo20170227.png 10:43:30 88 ms 13.23 KB Compl...
200 GET home.cnblogs.com /css/reset.min.css?v=NNKSv2cA90mrhu98reN... 10:43:30 42 ms 1.89 KB Compl...
200 GET home.cnblogs.com /css/home_common.css?v=8h9V8hkCCajAlYCQ... 10:43:30 43 ms 7.68 KB Compl...
200 GET home.cnblogs.com /css/follower.min.css?v=E-HDunl6YYlyha0XG2F... 10:43:30 39 ms 2.22 KB Compl...
图 6.17　博客园登录请求地址界面

在 Postman 请求头 Headers 中输入 Cookie 及其对应的值，如图 6.18 所示。

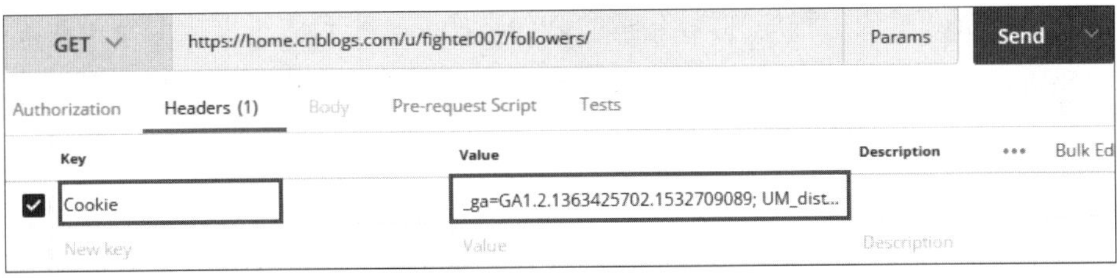
GET
https://home.cnblogs.com/u/fighter007/followers/
Params
Send
Authorization
Headers (1)
Body
Pre-request Script
Tests
Key
Value
Description
Bulk Ed
Cookie
_ga=GA1.2.1363425702.1532709089; UM_dist...
New key
Value
Description
图 6.18　请求头信息新增 Cookie 界面

单击“Send”按钮发送，查看运行结果：

```
<!DOCTYPE html>
<html>
<head>
    <meta charset="utf-8" />
    <meta name="viewport" content="width=device-width,initial-scale=1,maximum-
scale=1.0,minimum-scale=1.0,user-scalable=no" />
    <title> 我的粉丝 - 我的园子 - 博客园 </title>
    <link href="//common.cnblogs.com/favicon.ico" rel="shortcut icon" />
    <linkrel="stylesheet"href="/css/reset.min.css?v=NNKSv2cA90mrhu98reNDRA-Me_
BeHP9fhzvfTv39pb0" />
```

```
                <linkrel="stylesheet"href="/css/home_common.css?v=8h9V8hkCCajAlYCQTOkqgF-
CYDhdI0sknRaUIsm8dI8"/><em><a href="https://q.cnblogs.com/q/new" target="_self">提问
</a></em><img src="//common.cnblogs.com/images/ico_question.gif"alt=""><aid="app_q"
href="https://q.cnblogs.com/">博问 </a>
            </li>
            <li>
                <em><a href="#" target="_blank" onclick="AddToWz();return false;">
添加 </a></em><img src="//common.cnblogs.com/images/ico_bookmark.gif" alt=""><a id=
"app_wz" href="//wz.cnblogs.com/">收藏 </a>
            </li>
            <li>
                <em><a href="//job.cnblogs.com/admin/" target="_blank">发布 </a></em>
<img src="//common.cnblogs.com/images/ico_job.gif" alt=""><a id="app_job" href=
"/jobs/" >招聘 </a>
```

6.4 基于 Web Services 接口实战

Web Services 就是一个应用程序，它向外界暴露出一个能通过 Web 进行调用的 API，即能用编程的方法通过 Web 来调用这个应用程序。把调用这个 Web Services 的应用程序称为客户端，而把提供这个 Web Services 的应用程序称为服务端。Web Services 平台的三大技术分别是 XML+XSD、SOAP 和 WSDL。

（1）XML+XSD：定义了一套标准的数据类型，并给出了一种语言来扩展这套数据类型。

（2）SOAP：Web Services 通过 HTTP 发送请求和接收结果时，发送的请求内容和结果内容都采用 XML 格式封装，并增加了一些特定的 HTTP 消息头，以说明 HTTP 消息的内容格式，这些特定的 HTTP 消息头和 XML 内容格式就是 SOAP。

（3）WSDL：基于 XML 的语言，用于描述 Web Services 及其函数、参数和返回值。它是 Web Services 客户端和服务器端都能理解的标准格式。

6.4.1 Web Services 接口示例

Web Services 是建立可互操作的分布式应用程序的新平台，还是一套标准。它定义了应用程序如何在 Web 上实现互操作性，用户可以用任何语言，在任意平台上编写 Web Services，只要可以通过 Web Services 标准对这些服务进行查询和访问。本小节使用天气预报 Web Services 接口进行实战演示。

访问 http://www.webxml.com.cn/zh_cn/web_services.aspx 网站，其中有很多 Web Services 的接口案例。首先构建一个测试项目，在 Collections 下新增 WeatherPro 文件夹，在 WeatherPro 文件夹下

创建一个二级目录并命名为"getRegionProvince",如图 6.19 所示。

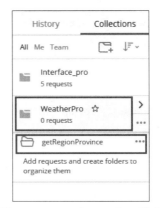

图 6.19　getRegionProvince 界面

新增获取省份接口,地址如下:http://ws.webxml.com.cn/WebServices/WeatherWS.asmx/getRegionProvince?,请求方法为 GET,将接口信息填入 Postman 工具中并发送请求,如图 6.20 所示。

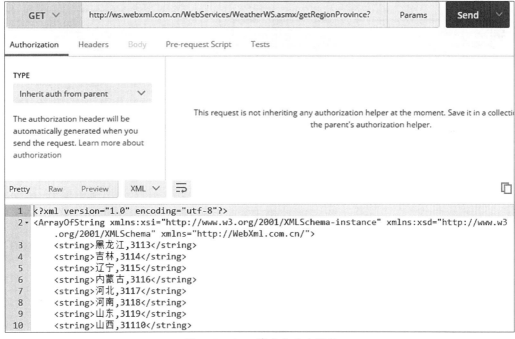

图 6.20　GET 请求和响应界面

将调试好的接口用例保存到创建的 getRegionProvince 文件夹中,并命名为"获取省份 ID 信息"。继续新增一个 POST 接口(获取城市 ID 接口),通过"获取省份 ID 信息"接口可以得出广东省对应的编码是 31124。

示例如下:

```
POST /WebServices/WeatherWS.asmx/getSupportCityString HTTP/1.1
Host: ws.webxml.com.cn
Content-Type: application/x-www-form-urlencoded
Content-Length: length
theRegionCode = string
```

请求方式为 POST，请求地址为 http://ws.webxml.com.cn/WebServices/WeatherWS.asmx/getSupportCityString，请求头信息为 Content-Type:application/x-www-form-urlencoded，请求参数为 theRegionCode=string（string 填写广东省份编码 31124），将接口信息填入 Postman 工具中并发送请求，如图 6.21 所示。

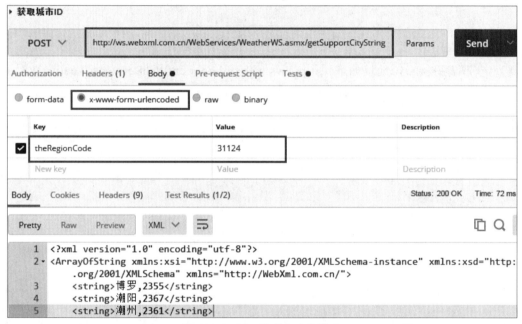

图 6.21　POST 请求和响应界面

在 WeatherPro 项目文件夹下新建二级目录并命名为"getSupportCity"，将创建好的 POST 请求保存到新建的二级目录 getSupportCity 中并命名为"获取城市 ID 信息"。

6.4.2　引用 JavaScript 断言策略

在接口测试中断言非常重要，没有断言就无法证明接口是否成功。Postman 工具的断言方法相比其他接口测试工具更复杂一些，是通过 JavaScript 脚本来实现的。本小节将结合 JavaScript 语法来演示 Postman 中的多种断言策略。

继续对 getRegionProvince 文件夹下的"获取省份 ID 信息"接口增加断言信息，如图 6.22 所示。

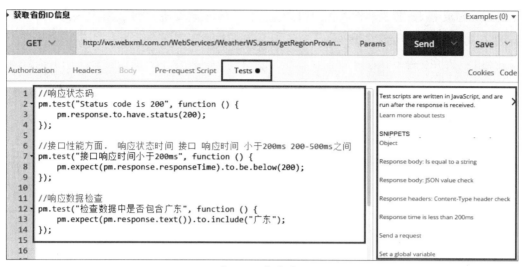

图 6.22 断言界面

如图 6.22 所示，左侧是编写断言位置，右侧提供了很多 JavaScript 函数。针对"获取省份 ID 信息"接口增加了 3 种常用的断言策略，分别是响应状态码断言、响应状态时间断言和响应数据检查断言。

增加断言后，再次运行接口，如图 6.23 所示。

图 6.23 断言结果界面

通过"Test Results"选项卡可以看出断言的 3 个结果都是 PASS 状态。一般来说，测试接口的响应时间是接口性能的主要考虑因素，对于单个接口响应时间，一般不超过 200ms 都是正常的。

6.4.3 解决动态参数获取

除可以获取响应信息中部分文本断言外，还可以通过获取 XML 数据 string 标签下博罗对应的编码 2335 做断言，这种断言处理相对麻烦，如图 6.24 所示。

```
<?xml version="1.0" encoding="utf-8"?>
<ArrayOfString xmlns:xsi="http://www.w3.org/2001/XMLSchema-instance" xmlns:xsd="http://www.w3.org/2001
/XMLSchema" xmlns="http://WebXml.com.cn/">
    <string>博罗,2355</string>
    <string>潮阳,2367</string>
    <string>潮州,2361</string>
```

图 6.24 获取城市名称和 ID 界面

先获取响应的 XML 数据，然后对 XML 数据进行拆分，最后获取拆分后右半部分即可。但是在 JavaScript 函数中没有提供解析 XML 数据的函数，所以只能先将 XML 数据转换为 JSON 数据，然后再进行解析和拆分，如图 6.25 所示。

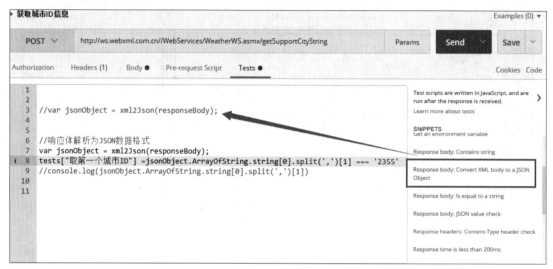

图 6.25　XML 数据转换为 JSON 数据界面

xml2Json() 方法用于将 XML 数据转换为 JSON 数据；jsonObject 是转换后定义的新对象；tests["取第一个城市ID"] 是自定义的断言信息，其中 tests 是 Postman 统一的断言格式；ArrayOfString.string[0] 表示获取 XML 数据中根节点下的第一个 string 标签；split() 方法用于将整个"博罗，2355"拆分成两部分；[1] 表示获取右半部分，即 2355。

查看运行结果，如图 6.26 所示。

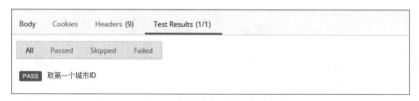

图 6.26　数据转换断言结果界面

6.4.4　解决上下游动态参数依赖

新增获取城市天气预报的数据接口。需要先传入城市 ID 和地区 ID（地区 ID 可以默认不填写）参数，然后返回数据，即一维字符串数组。接口信息如下：

```
POST /WebServices/WeatherWS.asmx/getWeather HTTP/1.1
Host: ws.webxml.com.cn
Content-Type: application/x-www-form-urlencoded
Content-Length: length
```

```
theCityCode = string&theUserID = string
```

将接口信息填入 Postman 工具中，如图 6.27 所示。

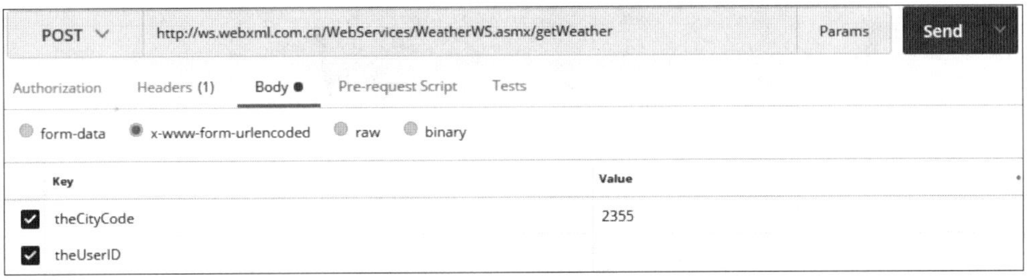

图 6.27　请求界面

2355 是博罗对应的城市编码，单击"Send"按钮发送，查看运行结果：

```
<?xml version="1.0" encoding="utf-8"?>
<ArrayOfString xmlns:xsi="http://www.w3.org/2001/XMLSchema-instance"
xmlns:xsd="http://www.w3.org/2001/XMLSchema" xmlns="http://WebXml.com.cn/">
    <string>广东 惠州 </string>
    <string>博罗 </string>
    <string>2355</string>
    <string>2018/12/24 20:16:22</string>
    <string>今日天气实况：气温：14℃；风向/风力：西北风 1级；湿度：98%</string>
    <string>紫外线强度：最弱。空气质量：良。</string>
    <string>紫外线指数：最弱，辐射弱，涂擦SPF8-12防晒护肤品。
健臻·血糖指数：不易波动，天气条件不易引起血糖波动。
穿衣指数：较冷，建议穿厚外套加毛衣等服装。
洗车指数：不宜，有雨，雨水和泥水会弄脏爱车。
空气污染指数：良，气象条件有利于空气污染物扩散。
```

将该接口保存在 getSupportCity 文件夹下并命名为"获取城市天气信息"。

通过上面的接口案例不难发现，"获取城市天气信息"接口（获取城市编码 ID）依赖于上一个接口（获取城市 ID 信息）的返回。目前在 Postman 中参数是固定，需要将其变为动态参数。

通过设置全局环境变量策略来解决动态参数的依赖问题，使用全局变量的好处是所有项目（Collections）都可以共享使用。在 Tests 中修改如下代码：

```
var jsonObject = xml2Json(responseBody);
var code = jsonObject.ArrayOfString.string[0].split(',')[1]
tests["get one CityCode"] = code === '2355';
// 将城市编码 ID 设置到全局环境变量中
//pm.globals.set("variable_key","variable_value");
pm.globals.set("citycode",code);
```

citycode 是提供下一个接口引用自定义的变量名，code 是获取的当前城市 ID 信息接口（博罗对应的城市编码 2355）。单击"Send"按钮，可以查看全局环境变量设置，如图 6.28 所示。

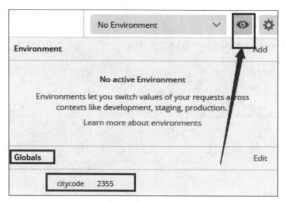

图 6.28 查看全局变量界面

将"获取城市天气信息"原接口请求参数 2355 改为 {{citycode}},再次运行接口并查看响应结果,如图 6.29 所示。

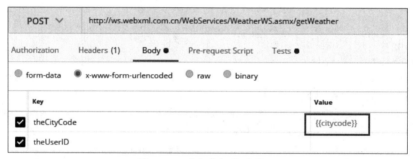

图 6.29 动态参数修改界面

本案例中动态参数依赖的处理思路是:将上一个接口(获取城市编码 ID)返回的城市编码 ID 设置到 Postman 全局环境变量中,当下一个接口(获取城市天气信息)需要用到时,可以直接引用全局变量的值。这就完成了接口和接口之间的关联调用和值传递。

6.4.5 Collections 公共数据分离

在前面的介绍中,所有的接口测试用例都是针对同一个服务器地址而言的,后期的维护比较麻烦。例如,写了 200 个接口用例,每个接口都是用同一个服务器地址或端口号,当服务器地址或端口号变化时,所有的用例都要修改对应的地址和端口,这样就比较麻烦。本小节进行 Postman 公共数据分离演示,以解决上述问题。

WeatherPro 项目下的所有接口使用的都是同一个服务器地址,可以在该项目下对服务器进行初始化操作,方便该项目下的所有用例调用。在 wearther 项目文件夹上右击,在弹出的快捷菜单中选择"Edit"选项,切换到"Varables"选项卡,然后输入公共数据服务器地址,如图 6.30 所示。

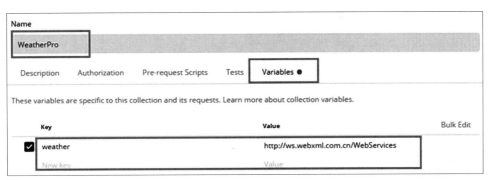

图 6.30　设置默认变量界面

分别修改"获取城市 ID 信息""获取城市天气信息""获取省份 ID 信息"接口请求地址参数，如图 6.31 所示。

图 6.31　请求地址修改界面

6.4.6　批量运行 Collections

当编写了大批量的接口自动化测试用例时，可以使用 Postman 提供的 Runner 插件来一次性地运行 Collections 项目下的所有测试用例。单击 Postman 工具栏的 "Runner" 按钮，打开如图 6.32 所示界面。

- Environment：如果接口中用到系统环境变量，需要切换对应的环境变量。
- Iterations：表示循环的次数。
- Delay：每个接口和接口之间的等待间隔。
- Log Responses：输出的日志文件。
- Data：外部数据文件。
- Run WeatherPro：运行 WeatherPro 测试项目。

图 6.32 Runner 运行设置界面

运行结果如图 6.33 所示。

图 6.33 运行结果界面

图 6.33 中显示了 3 条测试用例，第一条用例有 3 个断言，第二条用例有 1 个断言，第三条用例有 1 个断言。

6.4.7 使用 Newman+Jenkins 构建接口自动化任务

结合 Newman+Jenkins 可以实现真正的无人值守接口自动化测试任务，同时还可以利用 Jenkins 发送邮件告知测试结果，非常方便。

Newman 可以通过命令行方式执行 Postman 导出的测试脚本。使用 Newman 前必须要安装 Node.js 环境。访问 Node.js 官网（http://nodejs.cn/download）下载 Node.js，下载时注意区分操作系统版本。

注意：下载完成后需要将 Node.js 目录下 npm 所在的目录路径配置到系统的环境变量中。

打开 cmd 命令提示符界面，并进行以下验证：

```
C:\Users\23939>node -v
v7.7.1
```

（1）在线安装 Newman：

```
C:\Users\23939>npm  install -g  newman
```

（2）验证 Newman：

```
C:\Users\23939>newman -v
4.2.3
```

（3）安装 newman-reporter-html：

```
C:\Users\23939>npm install -g newman-reporter-html
npm WARN newman-reporter-html@1.0.2 requires a peer of newman@4 but none was installed.
```

（4）导出测试脚本。右击 WeatherPro 测试项目，在弹出的快捷菜单中选择"export"选项，打开如图 6.34 所示界面。

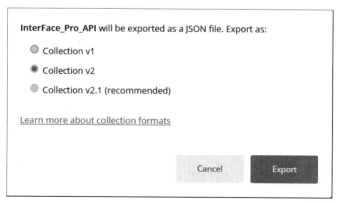

图 6.34　Postman 脚本导出界面

选择需要导出的测试脚本，然后单击"Export"按钮，即可导出 Globals 环境变量，如图 6.35

所示。

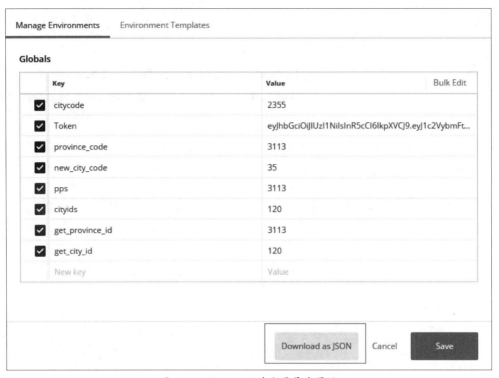

图 6.35　Globals 环境变量导出界面

（5）Newman 命令行运行 postman 脚本：

```
C:\Users\23939>newman run C:\Users\23939\Desktop\WeatherPro.postman_collection.json
-g C:\Users\23939\Desktop\globals.postman_globals.json --reporters html --reporter-
html-export D:\demo.html
newman: Newman v4 deprecates support for the v1 collection format
   Use the Postman Native app to export collections in the v2 format
```

命令解析：

```
newman run WeatherPro.postman_collection.json -g globals.postman_globals.json
--reporters  --reporter-html-export D:/PostmanScript.html
```

- WeatherPro.postman_collection.json：指定 Postman 运行测试脚本。
- globals.postman_globals.json：指定 (-g)Globals 环境变量脚本。
- --reporter-html-export D:/demo.html：表示将测试结果输出在 D 盘下。

（6）Jenkins 平台运行 Postman 测试脚本。

首先，在 Jenkins 机器上安装 Newman 环境。在 Jenkins 主界面选择"系统管理"→"节点管理"选项，新增 nodejs 和 npm 环境变量，单击"保存"按钮，如图 6.36 所示。

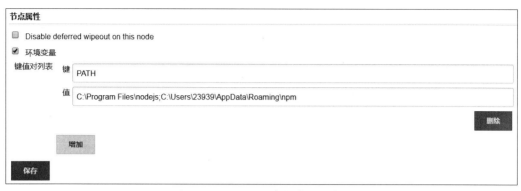

图 6.36　Jenkins 环境变量设置界面

注意：npm 的环境变量可以通过在 cmd 命令提示符界面中输入 "npm config get prefix" 来查看。

其次，新建一个自由风格的任务；切换到"构建"选项，在"Execute Windows batch command"界面中的"命令"文本框中输入 Newman 运行命令，如图 6.37 所示。

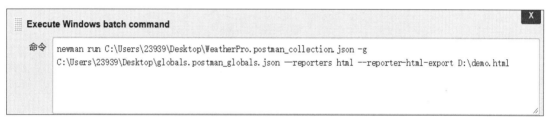

图 6.37　Jenkins 构建设置界面

查看控制台输出结果，如图 6.38 所示。

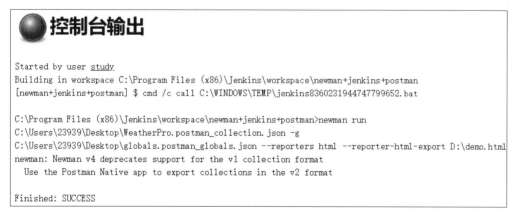

图 6.38　控制台输出结果界面

查看生成的本地 HTML 测试报告，如图 6.39 和图 6.40 所示。

```
Newman Report
Collection                              WeatherPro
Time                                    Sat Dec 22 2018 18:10:26 GMT+0800 (中国标准时间)
Exported with                           Newman v4.2.3

                                        Total                                                   Failed
Iterations                              1                                                       0
Requests                                3                                                       0
Prerequest Scripts                      4                                                       0
Test Scripts                            7                                                       0
Assertions                              7                                                       0

Total run duration                                                      510ms
Total data received                                                     3.07KB (approx)
Average response time                                                   81ms

Total Failures                          0
```

图 6.39　测试报告汇总界面

```
getRegionProvince

  getProvinceID

    Method                              GET
    URL                                 http://ws.webxml.com.cn/WebServices/WeatherWS.asmx/getRegionProvince

    Mean time per request               131ms
    Mean size per request               1.33KB

    Total passed tests                  3
    Total failed tests                  0

    Status code                         200

    Tests
```

Name	Pass count	Fail count
Status code is 200	1	0
Response time is less than 200ms	1	0
Body matches string	1	0

图 6.40　测试用例细节界面

第7章
Python接口自动化测试实战

前面介绍了接口测试工具 Postman 在接口测试中的应用。本章使用 Python+Requests 方式来进行接口自动化测试实战，从 Requests 常用示例（GET、POST、JSON、Requests Headers、Response 等）入手，围绕 Cookie、Session、Token 等案例进行实战演示，最后整合 DDT 数据驱动进行接口自动化测试实战。在本章最后会给读者演示接口框架的设计和开发思路。

7.1 安装 Requests 库

使用 Python+Requests 方式进行接口测试会更加灵活，代码也易于维护和扩展。安装和使用 Requests 进行接口测试非常简单，本节进行 Requests 库的安装及使用。

1. 在线安装方式

打开 cmd 命令提示符界面，输入 "pip install requests" 进行在线安装，示例如下：

```
C:\Users\Administrator>pip install requests
Collecting requests
  Downloading https://files.pythonhosted.org/packages/6d/e3/20f3d364d6c8e5d2353c
62a66668eb189166f08e863c9900e11c0286b84b/requests-2.21.0-py2.py3-none-any.whl
(56kB)
    60% |████████████████████████████     | 40kB 126kB/s e
    99% |████████████████████████████████ | 153kB
    69% |████████████████████████████     | 40kB 2.9MB/s e
    86% |████████████████████████████     | 51kB 2.8M
   110% |████████████████████████████████ | 61kB
 2.0MB/s
Installing collected packages: chardet,certifi,idna,requests
Successfully installed certifi-2018.11.29 chardet-3.0.4 idna-2.8 requests-2.21.0
```

2. 离线安装方式

下载最新的 Requests 安装包，打开网址 https://pypi.org/project/requests/#files 下载 .tar.gz 格式的离线包（图 7.1），解压到指定目录下，然后输入 "python setup.py install"，即可完成离线安装。

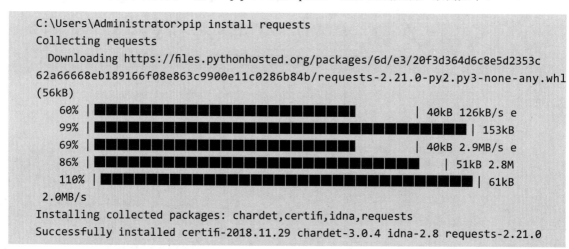

图 7.1 Requests 安装包界面

7.2 Requests 发送请求与参数传递

本节介绍在接口测试中使用 Requests 发送 GET、POST、JSON 请求，以及定制请求头信息处理的示例。

7.2.1 定义请求样式

使用 Requests 发送网络请求非常简单。导入 Requests 库，统一的接口样式发送格式如下。

（1）导入 Requests 库：

```
Import requests
```

（2）发送 GET 请求：

```
r = requests.get('https://api.github.com/events')
```

（3）发送 POST 请求：

```
requests.post("http://httpbin.org/post")
```

（4）发送 DELETE 请求：

```
requests.delete("http://httpbin.org/delete")
```

（5）发送 HEAD 请求：

```
requests.head("http://httpbin.org/get")
```

（6）发送 OPTIONS 请求：

```
requests.options("http://httpbin.org/get")
```

（7）发送 PUT 请求：

```
requests.put("http://httpbin.org/put")
```

7.2.2 发送 GET 请求

Requests 库发送 GET 请求的方式特别简单，可以携带参数传递，也可以不携带参数传递。先介绍一个不带参数的示例，代码如下：

```
import requests
```

```
response = requests.get('http://www.baidu.com')
print(r.url)
```

response 表示服务器响应的对象,这里可以获取所有想要的信息。输出结果如下:

```
http://www.baidu.com/
```

Requests 允许使用 params 关键字参数,以一个字符串字典来提供这些参数。例如,想传递 wd=python 这样的参数,可以使用如下代码:

```
import requests
r = requests.get('http://www.baidu.com/s?',params={'wd':'python'})
print(r.url)
```

输出结果如下:

```
http://www.baidu.com/s?wd=python
```

从结果上看,请求的参数被传入 URL 地址中。

7.2.3 发送 Form 表单

HTTP 规定 POST 提交的数据必须放在消息主体中,但是协议并没有规定必须使用何种编码方式。服务端通常是根据请求头中的 Content-Type 字段来获知请求中的消息主体是用何种方式进行编码,再对消息主体进行解析。常见的几种编码方式如下。

(1) application/x-www-form-urlencoded:以 Form 表单形式提交数据。提交的数据按照 key1=val1&key2=val2 的方式进行编码,key 和 val 都进行了 URL 转码。大部分服务端语言都支持这种方式。

(2) application/json:以 JSON 字符串方式提交数据。

(3) text/xml:以 XML 文档字符串方式提交数据。

(4) multipart/form-data:一般用来上传文件。

以禅道为例,使用 Charles 抓包来分析登录 POST 请求,如图 7.2 所示。

示例如下:

```
import requests
payload = {'account':'admin',
           'password':"123456",
           'referer':"http://192.168.0.162:4444/zentao/www/my/"}
r = requests.post('http://192.168.1.116:8011/zentao/www/user-login.html',
                   data=payload,
                   headers={"Content-Type":"application/x-www-form-urlencoded"})
print(r.text)
```

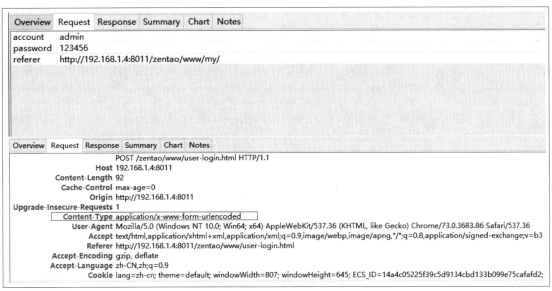

图 7.2　禅道登录请求头界面

payload 是将登录的请求参数构造为一个字典传递给 data 参数；headers 表示请求头信息，Content-Type 的作用是指明请求的正文类型是 application/x-www-form-urlencoded 格式，r.text 方法用于将服务器响应结果解析成文本字符串的 HTML 格式。响应结果如下：

```
<html><meta charset='utf-8'/><style>body{background:white}</style><script>parent.
location='http://192.168.0.162:4444/zentao/www/my/';
</script>
```

7.2.4　发送 XML 数据

使用 Requets 库中的 POST 请求发送标准的 XML 文档数据，示例如下：

```
import requests
url = 'http://ws.webxml.com.cn/WebServices/WeatherWS.asmx'
payload = '''<?xml version="1.0" encoding="utf-8"?>
<soap:Envelope xmlns:xsi="http://www.w3.org/2001/XMLSchema-instance"
               xmlns:xsd="http://www.w3.org/2001/XMLSchema"
               xmlns:soap="http://schemas.xmlsoap.org/soap/envelope/">
  <soap:Body>
    <getSupportCityString xmlns="http://WebXml.com.cn/">
      <theRegionCode>3114</theRegionCode>
    </getSupportCityString>
  </soap:Body>
</soap:Envelope>'''
r = requests.post(url=url,data=payload,headers={"Content-Type":"text/xml"})
print(r.text)
```

payload 对象是发送标准的 XML 字符串文档，"Content-Type":"text/xml" 表示指明请求的正文类型是 XML 文档格式。输出结果如下：

```
<?xml version="1.0" encoding="utf-8"?><soap:Envelope xmlns:soap="http://schemas.
xmlsoap.org/soap/envelope/" xmlns:xsi="http://www.w3.org/2001/XMLSchema-instance"
xmlns:xsd="http://www.w3.org/2001/XMLSchema"><soap:Body><getSupportCityString
Response xmlns="http://WebXml.com.cn/"><getSupportCityStringResult><string>安图
,658</string><string>白城,116</string><string>白山,636</string><string>长白,
640</string><string>长春,650</string><string>长岭,622</string><string>大安,
111</string><string>德惠,625</string><string>东丰,683</string><string>东岗,
690</string><string>敦化,656</string><string>扶余,623</string><string>公主岭
,648</string><string>和龙,692</string><string>桦甸,686</string><string>珲春,
695</string><string>辉南,688</string><string>吉林,655</string><string>集安,
639</string><string>蛟河,656</string><string>靖宇,689</string><string>九台,
626</string><string>梨树,646</string><string>辽源,682</string><string>临江,
638</string><string>柳河,686</string><string>龙井,694</string><string>梅河口,
685</string><string>农安,624</string><string>磐石,684</string><string>前郭,
114</string><string>乾安,113</string><string>舒兰,628</string><string>双辽,
645</string><string>双阳,652</string><string>四平,649</string><string>松原,
112</string><string>洮南,119</string><string>通化,635</string><string>通化县,
634</string><string>通榆,620</string><string>图们,696</string><string>汪清,
660</string><string>延吉,696</string><string>伊通,651</string><string>永吉,
654</string><string>榆树,626</string><string>镇赉,111</string><string>东辽,
3566</string><string>抚松,3566</string></getSupportCityStringResult></getSupportCity
StringResponse></soap:Body></soap:Envelope>
```

7.2.5 发送 JSON 请求

有些接口请求的正文类型是 JSON 字符串，可以使用 Requests 库发送 POST 请求处理 JSON 字符串文本，示例如下：

```
import requests,json
url = 'http://183.62.166.42:11125/CyzgMobileConfigService/GetDataInfo'
payload = {'CommandCode': 'GetAllCityData',
           'Marker': '1482638389646',"TransferData": "{\'CityId\':4565110}"}
r = requests.post(url=url,json=payload,headers={'Content-Type': 'application/json'})
print(r.text)
```

'Content-Type':'application/json' 表示指明请求的正文类型是 JSON 字符串文本；json=payload 表示 JSON 会自动把后面的 payload 字典拼接为标准的 JSON 字符串，然后放在 POST 请求正文中进行提交。输出结果如下：

```
{"ErrorInfo":" 城市信息获取成功 ","ResultCode":11001,"ResultInfo":"[{\"province_
id\":\"430000\",\"province_name\":\"湖南省 \",\"cities\":[{\"city_id\":\"430110\",
\"city_name\":\" 湖南省长沙市 \",\"is_appoint\":\"0\",\"isNotCheckFace\":\"0\",
```

```
\"appoint_Url\":\"\",\"examOverValue\":0,\"mnksOverValue\":60,\"dycsOver
Value\":60,\"server_Url\":\"http:\/\/node1.m.jtcyzg.cn:11150\/Changsha\/
CyzgMobileService.svc\/\",\"exam_Url\":\"\",\"fileLoad_Url\":\"http:\/\/jxjy-app.
oss-cn-shenzhen.aliyuncs.com\/app_setup\/JXJY_Phone_v1.1.2.apk\",\"mrServer_Url
\":\"http:\/\/166.66.189.445:11118\/CyzgTaxiService.svc\/\",\"mrFileLoad_Url\":
\"http:\/\/192.168.50.44:11286\/ChapterFlvHandler.ashx\/\",\"helpVedio_Url\":\
"http:\/\/192.168.50.30:11026\/ChapterFlvHandler.ashx\/HelpVedio\/0.mp4\",\"IsNee]}
```

7.3 处理 Token

Token 翻译过来是 "令牌"，也可以理解为身份认证。目前大部分互联网产品都是使用基于 Token 的身份认证方法。Token 是服务端生成的一串字符串，当第一次登录后，服务器会签发一个 Token，然后将此 Token 返回给客户端，客户端拿到 Token 以后可以把它存储起来，如放在 Cookie 和 Headers 中。客户端每次向服务端请求资源时，都需要带上服务端签发的 Token，服务端收到请求后需验证客户端请求里面携带的 Token 是否和签发时一致，如果验证成功就向客户端返回请求的数据。

示例如下：

```python
import requests
url = 'http://212.123.456.123:1112/login'
data = {'username':'admin','password':'admin'}
response = requests.post(url=url,json=data,
                        headers={'Content-Type':"application/json"})
print(response.json())
```

json=data 表示将 data 字典编码为 JSON 字符串并放在 POST 请求体正文，'Content-Type':"application/json" 表示指明请求的正文类型是 JSON 字符串文本。响应结果信息如下：

```
{'username':'admin',
 'token':'eyJhbGciOiJIUzI1NiIsInR5cCI6IkpXVCJ9.eyJ1c2VybmFtZSI6ImFkbWl
 uIiwiaWQiOjEyMTYsImlhdCI6MTU0NTgwNzc4NywiZXhwIjoxNjMyMjA3uunmjk.WVpbX_
 qFVzsfsc43PwB6pQUrXv0qCOMzNDFU_o4GY8A','id': 1216}
```

查看响应结果服务器返回给客户端 Token 信息，当访问下一个接口（获取所有任务接口）时，需要带上 Token 信息才可以访问。继续新增查看任务的接口信息，示例如下：

```python
import requests
import unittest
class ApiTest(unittest.TestCase):
    @classmethod
```

```python
    def setUpClass(cls):
        cls.url = 'http://212.123.456.123:1112/login'
@classmethod
    def tearDownClass(cls):
        pass
    def write_token(self):
        '''将 Token 信息写入到 token.md 文件中'''
        data = {'username':'admin','password':'admin'}
        response = requests.post(url=self.url,json=data,
                                headers={'Content-Type':"application/json"})
        with open('token.md','w') as f:
            f.write(response.json()['token'])
    def read_Token(self):
        '''读取 token.md 文件中的 Token 信息'''
        with open('token.md','r') as f:
            return f.read()
    def test_getApiTask(self):
        '''验证获取所有任务接口'''
        response = requests.get(url=self.url+'/api/tasks',headers=self.readToken())
        self.assertEqual(response.status_code,200)
if __name__ == '__main__':
    unittest.main(verbosity=2)
```

write_token() 方法的作用是将登录后获取的 Token 写入 token.md 文件中；read_Token() 方法用于读取 token.md 文件中的 Token，并将读取的 Token 作为请求参数传入下一个接口（获取所有任务接口）的请求头信息 Headers 中，这样就可以获取所有任务接口信息。响应代码如下：

```
[
  {
    "id": 960,
    "title": " 老鹿的菜园子 ",
    "desc": "https://www.cnblogs.com/fighter006",
    "done": false,
    "createdAt": "2018-12-25T15:13:44.000Z",
    "updatedAt": "2018-12-25T15:13:44.000Z"
  },
  {
    "id": 959,
    "title": "https://www.cnblogs.com/fighter006",
    "desc": "",
    "done": false,
    "createdAt": "2018-12-25T06:00:12.000Z",
    "updatedAt": "2018-12-25T06:00:12.000Z"
  }
]
```

7.4 处理 Cookie

关于 Cookie 的工作原理在 4.7.1 小节有相关介绍。本节演示 Requests 库在接口测试过程中对 Cookie 的处理。以禅道为例，不登录禅道则无法查看 Bug 单。如图 7.3 所示，抓取禅道登录接口。

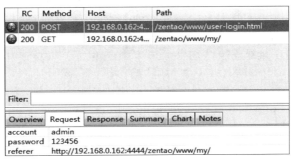

图 7.3　请求参数界面

示例如下：

```
def login():
    '''登录禅道获取登录后的session'''
    r = requests.post('url=http://192.168.0.162:4444/zentao/www/user-login.html',
                    data={'account':'admin',
                          'password':'123456',
                          'referer':"http://192.168.0.162:4444/zentao/www/my/"})
    return r.cookies
```

将 Charles 抓取的请求参数 url、account、password 和 referer 分别填入 data 字典中，login() 函数中 r.cookies 方法的作用是获取登录禅道后服务器返回的 Session 信息。

```
def ViewBug():
    '''查看Bug单'''
    r = requests.get(url='http://192.168.0.162:4444/zentao/www/my-bug.html',
                    cookies=login())
    return r.text
print(ViewBug())
```

查看 Bug 单接口需要用到登录接口（login() 方法）返回的 Session 信息。cookies=login() 方法表示把登录成功后服务器返回的 Session 信息作为请求参数传入第二个接口（查看 Bug 接口）中。
响应信息如下：

```
<!DOCTYPE html>
<html lang='zh-cn'>
```

```html
<head>
  <meta charset='utf-8'>
  <meta http-equiv='X-UA-Compatible' content='IE=edge'>
  <meta name="renderer" content="webkit">
  <title> 我的地盘 :: 我的 Bug - 禅道 </title>
</head>
<body>
<script>
var config={"webRoot":"\/zentao\/www\/","cookieLife":30,"requestType":"PATH_INFO",
"pathType":"clean","requestFix":"-","moduleVar":"m","methodVar":"f","viewVar":"t",
"defaultView":"html","themeRoot":"\/zentao\/www\/theme\/","currentModule":"my",
"currentMethod":"bug","clientLang":"zh-cn","requiredFields":"","router":
"\/zentao\/www\/index.php\/my-bug.html","timeout":30000};
var lang={"submitting":"\u7a0d\u5019...","save":"\u4fdd\u5b58","timeout":"\u8fde\
u63a5\u8d85\u65f6\uff0c\u8bf7\u68c0\u67e5\u7f51\u7edc\u73af\u5883\uff0c\u6216\
u91cd\u8bd5\uff01"};
</script>
<ul class='nav'>
<li class='active'><a href='/zentao/www/my/' class='active' id='menumy'>
<i class="icon-home"></i><span> 我的地盘 </span></a></li>
<li><a href='/zentao/www/product/' id='menuproduct'> 产品 </a></li>
<li><a href='/zentao/www/project/' id='menuproject'> 项目 </a></li>
<li><a href='/zentao/www/qa/' id='menuqa'> 测试 </a></li>
<li><a href='/zentao/www/doc/' id='menudoc'> 文档 </a></li>
<li><a href='/zentao/www/report/' id='menureport'> 统计 </a></li>
<li><a href='/zentao/www/company/' id='menucompany'> 组织 </a></li>
<li><a href='/zentao/www/admin/' id='menuadmin'> 后台 </a></li>
</ul>
</body>
</html>
```

7.5 处理 Session

Session 表示会话对象，在 4.7.2 小节有相关介绍。本节使用 Requests 库来演示 Session 在接口自动化测试中的应用。

以人人网为例，先使用 Charles 抓包分析登录接口，如图 7.4 所示。

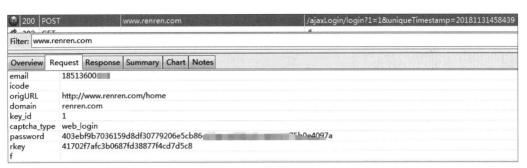

图 7.4 人人网 Request 界面

使用 Requests 实现 POST 请求登录过程，示例如下：

```
import requests
data = {
    "email":"18513600xxx",
    "icode":"",
    "origURL":"http://www.renren.com/home",
    "domain":"renren.com",
    "key_id":1,
    "captcha_type":"web_login",
    "password":"403ebf9eelri59d8df30669206e8885236klaece4884c0cca4642365b0e4096a",
    "rkey":"41602f6afc3b0686fd38866f4cd6d5c8",
    "f":""}
url = 'http://www.renren.com/ajaxLogin/login?1=1&uniqueTimestamp=20181131458439'
def RenrenLogin():
    ''' 登录接口 '''
    s = requests.Session()
    r = s.post(url=url,data=data)
    return s
```

将登录请求需要用到的参数分别填入 data 字典中。requests.Session() 方法表示实例化 Session 对象，保持后续发送的所有请求之间保持 Cookie。继续新增编辑个人信息接口，抓包信息如图 7.5 所示。

图 7.5 编辑个人信息 Request 界面

示例如下：

```python
def SetPersoninfo():
    ''' 设置个人信息 '''
    data = {"music":" 敢爱敢做 ",
            "interest":"https://www.cnblogs.com/fighter006",
            "book":" 墨菲定律 ",
            "movie":" 巨齿鲨 ",
            "game":" 不喜欢玩游戏 ",
            "comic":" 不看动漫  只学习 ",
            "sport":" 跑步  健身 ",
            "errorReturn":1,
            "submit":" 保存 ",
            "requestToken":"-1366234446",
            "_rtk":"69aa114b"}
    url = "http://www.renren.com/PersonalInfo.do?v=info_timeline"
    r = RenrenLogin().post(url=url,data=data)
    return r.text
print(SetPersoninfo())
```

将编辑个人信息接口请求参数填入 data 字典中。SetPersoninfo() 方法调用 RenrenLogin() 方法保证了请求之间的 Cookie 会话。查看修改个人信息后的页面效果，如图 7.6 所示。

图 7.6 人人网个人信息编辑界面

响应信息如下：

```html
<div class="info-section-head">
<h4>
<span class="icon-box"><img
<dl class="info">
<dt> 音乐 </dt>
<dd>
<a  stats="info_info"  href='http://browse.renren.com/searchEx.do?s=5&amp;p=%5B%6B%2
2t%22%3A%22fond%22%26%88%B1%E6%%22%6D%5D'> 敢爱敢做 </a></dd>
</dl>
<dl class="info">
<dt> 爱好 </dt>
<dd>
<a  stats="info_info"  href='http://browse.renren.com/searchEx.do?s=5&amp;p=%5B%6B%2
```

```html
2t%22%3A%22fond%22%2C%22name%22%3A%22inte%22%2C%22value%22%3A%22https%3A%2F%2Fwww.
cnblogs.co2%6D%5D'>https://www.cnblogs.com/fighter006</a></dd>
</dl>
<dl class="info">
<dt> 书籍 </dt>
<dd>
<a stats="info_info" href='http://browse.renren.com/searchEx.do?s=5&amp;p=%5B%6B%2
2t%22%3A%22fond%22%2AE9%A%E5%BE%8B%22%6D%2C%6B%22id%22%3A%223414%22%2C%22t%22%3A%2
2univ%22%6D%5D'> 墨菲定律 </a></dd>
</dl>
<dl class="info">
<dt> 电影 </dt>
<dd>
<a stats="info_info" href='http://browse.renren.com/searchEx.do?s=5&amp;p=%5B%6B%2
2t%22%3A%22fond%22%2B2%A8%22%6D%2C%6B%22id%22%3A%223414%22%2C%22t%22%3A%22univ%22
%6D%5D'> 巨齿鲨 </a></dd>
</dl>
<dl class="info">
<dt> 游戏 </dt>
<dd>
<a stats="info_info" href='http://browse.renren.com/searchEx.do?s=5&amp;p=%5B%6B%2
2t%22%3A%22fond%22%2'> 不喜欢玩游戏 </a></dd>
</dl>
<dl class="info">
<dt> 动漫 </dt>
<dd>
<a stats="info_info" href='http://browse.renren.com/searchEx.do?s=5&amp;p=%5B%6B%2
2t%22%3A%22fond%22%2'> 不看动漫 只学习 </a></dd>
</dl>
<dl class="info">
<dt> 运动 </dt>
<dd>
<a stats="info_info" href='http://browse.renren.com/searchEx.do?s=5&amp;p=%5B%6B%2
2t%22%3A%22fond%22%2'> 跑步 健身 </a></dd>
</dl>
</div>
```

7.6 处理超时等待

接口如果响应时间过长，可以用 timeout 设置超时时间。在指定时间内，接口未响应，此时就会抛出 requests.exceptions.ConnectTimeout 异常，示例如下：

```python
import requests   # 导入 Requests 库
response = requests.get(url='http://www.baidu.com',timeout=0.001)
print(response.text)
```

timeout=0.001 表示发送请求到接收响应信息时间间隔不超过 0.001s，如果超过设置的时长 0.001s，就会抛出 requests.exceptions.ConnectTimeout 异常。

7.7 Response 对象解析

当 Requests 发送请求时，服务器会返回一个 Response 对象，在 Response 中有很多重要信息，如状态码、响应头字段和原始响应体等。下面通过一个案例来演示 Response 对象中的重要信息，示例如下：

```python
import requests
r = requests.get('http://httpbin.org/get')
print('HTTP 状态码：',r.status_code)      # 响应状态码，常见的有 200、301、302、404、500
print('返回原始响应体 ',r.raw)
print(' 请求的响应体 ',r.content)    # 字节方式的响应体，会自动解码 gzip 和 deflate 压缩
print(' 响应内容 ',r.text)            # 字符串方式的响应体，会自动根据响应头部的字符编码进行解码
print(' 获取 headers ',r.headers)    # 以字典对象存储服务器响应头
print(r.json())          # Requests 中内置的 JSON 解码器，将响应结果转换为 JSON 字符串
print(r.raise_for_status())         # 失败请求（非 200 响应）抛出异常
```

响应结果如下：

```
HTTP 状态码： 200
返回原始响应体 <urllib3.response.HTTPResponse object at 0x00000156D89E6A58>
请求的响应体 b'{\n  "args": {},\n  "headers": {\n    "Accept": "*/*",\n    "Accept-Encoding": "gzip,deflate",\n    "Connection": "close",\n    "Host": "httpbin.org",\n    "User-Agent": "python-requests/2.19.1"\n  },\n  "origin": "183.12.113.251",\n  "url": "http://httpbin.org/get"\n}\n'
响应内容 {
  "args": {},
  "headers": {
    "Accept": "*/*",
    "Accept-Encoding": "gzip,deflate",
    "Connection": "close",
    "Host": "httpbin.org",
    "User-Agent": "python-requests/2.19.1"
  },
  "origin": "183.12.113.251",
  "url": "http://httpbin.org/get"
}
```

```
获取 headers {'Server': 'gunicorn/19.9.0','Date': 'Tue,25 Dec 2018 14:20:11 GMT',
'Content-Type': 'application/json','Content-Length': '266','Access-Control-Allow-
Origin': '*','Access-Control-Allow-Credentials': 'true','Via': '1.1 vegur','Proxy-
Connection': 'Keep-alive'}
{'args': {},'headers': {'Accept': '*/*','Accept-Encoding': 'gzip,deflate',
'Connection': 'close','Host': 'httpbin.org','User-Agent': 'python-
requests/2.19.1'},'origin': '183.12.113.251','url': 'http://httpbin.org/get'}
None
```

7.8 Requests 文件上传实战

实际接口测试过程中会经常遇到文件上传的场景，文件上传一般包含图片、视频和文档（Excel、CSV、记事本等）。本节使用 Requests 库来进行文件上传的实战演示。

以禅道为例，登录禅道后，在文档下单击创建文档的"+"，新建一个文档，然后提交。使用 Charles 抓包后，查看结果，如图 7.7 所示。

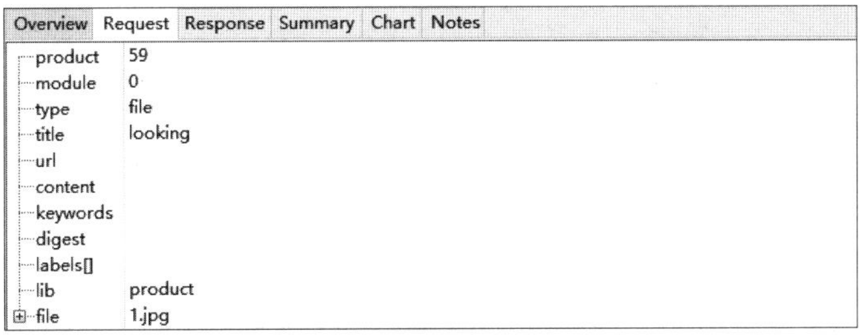

图 7.7 上传接口请求界面

如图 7.7 所示，在请求中可以查看提交的数据。选择"Response"→"Headers"选项，查看请求头中的 Content-Type 内容，如图 7.8 所示。

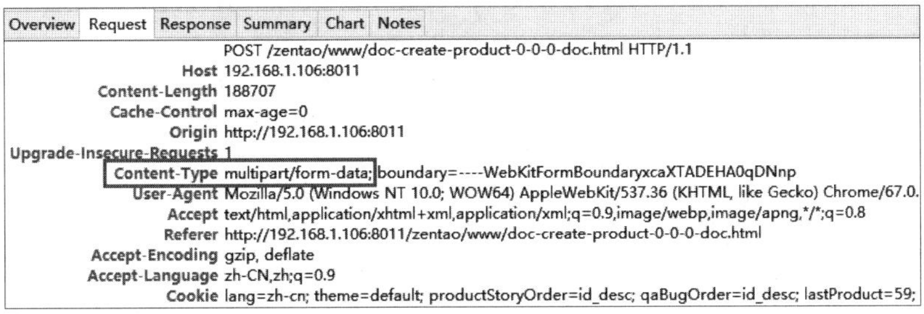

图 7.8 Content-Type 界面

multipart/from-data 表示指定的传输数据为二进制类型，如图片、mp3 和文件等，然后选择"Request"→"Raw"选项，查看文件的 Content-Type 内容，如图 7.9 所示。

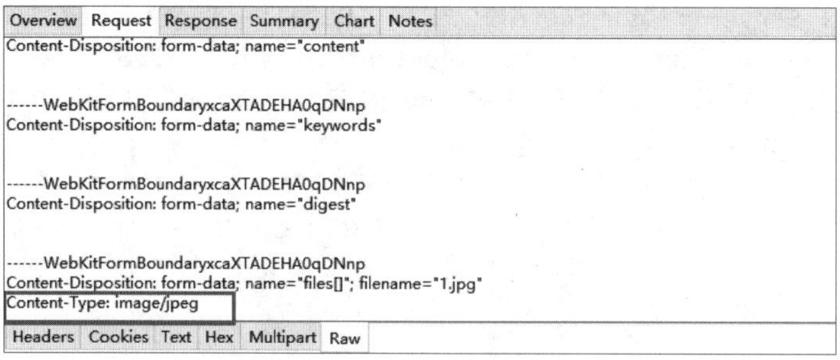

图 7.9 Raw 下的 Content-Type 界面

上传文件的这部分必须要指定文件的路径。关于文件的 Content-Type 类型，特别需要强调的是，file 其实就是请求参数中的 file，必须与请求参数中的参数名称一一对应，示例如下：

```
files = {"file":("jenkins.jpg",open("c:/jenkins.jpg","rb"),"image/jpeg",{})}
```

具体示例如下：

```python
def zentaoLogin():
    """ 登录请求 """
    data = {
        'account': 'admin',
        'password': '123456',
        'referer': 'http://192.168.1.106:8011/zentao/www/doc-browse-product-0-59-0.html'}
    return data
def login():
    """ 获取登录后返回Cookie"""
    r = requests.post(
        url='http://192.168.1.106:8011/zentao/www/user-login.html',
        data=zentaoLogin(),
        headers={'Content-Type': 'application/x-www-form-urlencoded'})
    return r.cookies
def uploadData():
    """ 接口参数 """
    data = {
        "product": "59",
        "module":  "0",
        'type': 'file',
        "title": 'looking1',
        'url': '',
        'content': '',
        'keywords': '',
        'digest':'',
        'labels[]':'',
        'lib':'product'
    }
    return data
def uploadFile():
```

```
""" 上传文件接口 """
    r = requests.post(
        url= 'http://192.168.1.106:8011/zentao/www/doc-create-product-0-0-0-doc.html',
        data=uploadData(),
        headers={'Conteny-Type': 'multipart/form-data'},
        files={"file": ("1.jpg",open(r"C:/1.jpg","rb"),"image/jpeg",{})},
        cookies=login())
    print(r.status_code)
    print(r.text)
uploadFile()
```

整个文件上传的流程是先登录系统，再带上登录系统返回的 Cookie 信息，然后进行文件上传操作。在 login() 方法中，r.cookies 方法表示获取登录后返回的 Cookie 信息，uploadFile() 方法表示上传文件接口，这里需要用对应入参（接口地址 url、请求参数 data、请求头 headers、文件类型 files 及 cookies）。

运行查看界面响应结果，如图 7.10 所示。

图 7.10　上传文件结果界面

Requests 常见异常

遇到网络问题（如 DNS 查询失败、拒绝连接等）时，Requests 会抛出一个 ConnectionError 异常；如果 HTTP 请求返回失败状态码，则 Response.raise_for_status() 会抛出一个 HTTPError 异常；如果请求超时，则抛出一个 Timeout 异常；如果请求超过了设定的最大重定向次数，则会抛出一个 TooManyRedirects 异常。所有 Requests 显式抛出的异常都继承自 requests.exceptions.RequestException 类。

序列化和反序列化

序列化可以理解为把 Python 的对象编码转换为 JSON 格式的字符串，反序列化可以理解为把 JSON 格式的字符串解码为 Python 数据对象。在 Python 标准库中，提供了 JSON 库与 pickle 库来处理序列化与反序列化。

JSON（JavaScript Object Notation）是一种轻量级基于文本的可读格式，采用完全独立于编程语言的文本格式来存储和表示数据。JSON 是 JS 对象的字符串表示法，使用文本表示一个 JS 对象的信息，本质是一个字符串。JSON 最常用的格式是对象的键值对。

JSON 示例如下：

```
{
  "name":"JSON 在线解析",
  "age":33,
  "msg":["datainfor","successful"],
  "regex":"https://www.cnblogs.com/fighter007"
};
```

JSON 库中提供的 dumps() 方法和 loads() 方法可以实现数据的序列化和反序列化。dumps() 方法是把 Python 数据类型转换为 JSON 相应的数据类型，loads() 方法可将 JSON 格式数据序列化为 Python 的相关数据类型（如列表、元组、字典等）。

（1）导入 JSON 库，查看 JSON 库中有哪些方法，示例如下：

```
>>> import json
>>> print(json.__all__)
['dump','dumps','load','loads','JSONDecoder','JSONEncoder']
>>>
```

（2）对 Python 中的字典进行序列化操作，示例如下：

```
import json    # 导入 JSON 库
dict1 = {'name':"fighter",'age':28,'address':'shenzhen'}
print('未序列化前的数据类型为：',type(dict1))
print('未序列化前的数据：',dict1)
str1 = json.dumps(dict1)    # 将 Python 数据对象序列化操作变成字符串
print('序列化后的数据类型为：',type(str1))
print('序列化后的数据为：',str1)
```

输出结果如下：

```
未序列化前的数据类型为：<class 'dict'>                    # ----------- 字典类型
未序列化前的数据：{'address': 'shenzhen','name': 'fighter','age': 28}
序列化后的数据类型为：<class 'str'>                       # ----------- 字符串类型
序列化后的数据为：{"address": "shenzhen","name": "fighter","age": 28}
```

序列化前 dict1 对象是字典类型，通过 JSON 库中的 dumps() 方法序列化后转换为 JSON 对象数据类型（str 字符串类型），从而实现了序列化操作。

（3）对序列化后的 JSON 对象进行反序列化操作，示例如下：

```
import json    # 导入 JSON 库
# 定义字典
```

```
dict1 = {'name':"fighter",
         'age':28,
         'address':'shenzhen'}
print('未序列化前的数据类型为:',type(dict1))
print('未序列化前的数据:',dict1)
# 对 Python 对象进行序列化操作
print('begin 对 Python 对象进行序列化操作 ------------>')
str1 = json.dumps(dict1)
print('序列化后的数据类型为:',type(str1))
print('序列化后的数据为:',str1)
# 对 str1 进行反序列化操作
print('begin 对 str1 对象进行反序列化操作 ------------>')
dict2 = json.loads(str1)
print('反序列化后的数据类型:',type(dict2))
print('反序列化后的数据:',dict2)
```

输出结果如下:

```
未序列化前的数据类型为: <class 'dict'>
未序列化前的数据: {'name': 'fighter','age': 28,'address': 'shenzhen'}
begin 对 Python 对象进行序列化操作 ------------>
序列化后的数据类型为: <class 'str'>
序列化后的数据为: {"name": "fighter","age": 28,"address": "shenzhen"}
begin 对 str1 对象进行反序列化操作 ------------>
反序列化后的数据类型: <class 'dict'>
反序列化后的数据: {'name': 'fighter','age': 28,'address': 'shenzhen'}
```

上述案例中,首先对 dict1 字典对象进行序列化操作转换为 str1(JSON 字符串),然后使用 JSON 库提供的 loads() 方法将序列化后的 str1(JSON 对象)转换为 Python 数据对象(字典类型)。

注意:JSON 库中的 dumps() 方法除可以序列化 Python 数据对象(字典)外,还可以序列化 Python 中的列表、元组等,其实现方法与字典的序列化和反序列化操作原理一致。

7.11 XML 与 JSON 数据之间的转换

Python 中 XML 和 JSON 两种数据格式是可以相互转换的,就像 JSON 格式转 Python 字典格式对象那样。XML 格式和 JSON 格式相互转换需要用到 xmltodict 库。下面安装 xmltodict 库,打开 cmd 命令提示符界面,输入"pip install xmltodict"进行在线安装,示例如下:

```
C:\Users\Administrator>pip install xmltodict
Collecting xmltodict
```

```
Downloading https://files.pythonhosted.org/packages/42/a9/6e99652c6bc619d19d58
cdd8c46560630eb5825d43a6e25db2e1d666ceb6/xmltodict-0.11.0-py2.py3-none-any.whl
Installing collected packages: xmltodict
Successfully installed xmltodict-0.11.0
```

示例如下:

```
import json
import xmltodict
# 定义 XML 转 JSON 的函数
def xmltojson(xmlstr):
    xmlparse = xmltodict.parse(xmlstr)    # parse 是 XML 解析器
    jsonstr = json.dumps(xmlparse,indent=2,sort_keys=True)
    print(jsonstr)
if __name__ == '__main__':
    xmlinfo = """<student>
        <bokeid>fighter006</bokeid>
        <bokeinfo>
            <cnbologsname>laolu</cnbologsname>
            <page>230</page>
        </bokeinfo>
        <date>
            <address>http://www.baidu.com</address>
            <title>python+ddt+requests</title>
        </date>
    </student>
    """
    xmltojson(xmlinfo)
```

parse() 方法是 XML 提供的解析器,dumps() 方法用来将 dict 转换为 JSON 格式。dumps() 方法中的 ident=2 表示输出的 JSON 格式缩进两个空格,起到美化作用;sort_keys=True 表示输出结果中的 JSON 键按照 ACSII 码排序显示。转换 JSON 后,输出结果如下:

```
{
  "student": {
    "bokeid": "fighter006",
    "bokeinfo": {
      "cnbologsname": "laolu",
      "page": "230"
    },
    "date": {
      "address": "http://www.baidu.com",
      "title": "python+ddt+requests"
    }
  }
}
```

同样地，JSON 数据也可以转换为 XML 数据，使用 xmltodict 库的 unparse() 方法将 JSON 格式转换为 XML 格式，示例如下：

```python
import xmltodict
def jsontoxml(jsonstr):
    xmlstr = xmltodict.unparse(jsonstr)
    print(xmlstr)
if __name__ == "__main__":
    json = {'cnblogs':
                {'title':
                    {'name': 'fighter006',
                     'address': 'https://www.cnblogs.com/fighter006/'},
                 'info': {'page': '230',
                          'updated': 'everyweekend'},
                 'job': 'tester'}
            }
    jsontoxml(json)
```

输出结果如下：

```
<cnblogs><job>tester</job><title><name>fighter006</name><address>https://www.cnblogs.com/fighter006/</address></title><info><page>230</page><updated>everyweekend</updated></info></cnblogs>
```

7.12 接口测试框架设计和开发

本节通过一个接口项目来演示接口测试框架的设计和开发思路。这里会涉及使用 unittest 单元测试框架对接口用例的设计和组织、重构 Requests 请求方法，重构操作 Excel 数据工具类，分离接口数据和接口用例（重点关注接口业务逻辑）及对接口动态参数赋值、调用、一次性解决上下游参数调用，以及接口运行过程中进行日志跟踪及最后的项目持续集成等，通过以上内容来一步步完善接口测试框架，给读者提供一个开发思路。

7.12.1 测试框架简介

整个接口测试框架的设计图如图 7.11 所示。

- Config 目录：存放配置文件，如数据库的端口、地址和邮件配置信息等。
- Data 目录：存放公共动态数据，如 Token、Excel、动态参数等。
- Log 目录：存放 Log 日志信息。

- Reports 目录：存放接口测试报告。
- TestCases 目录：存放接口测试案例。
- Utlis 目录：公共方法、自定义工具类所在目录。
- runMain.py 文件：项目运行的主程序文件。

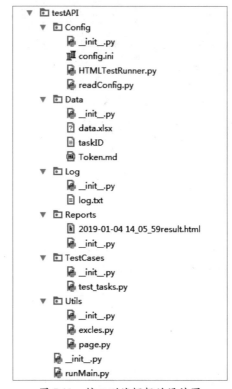

图 7.11　接口测试框架的设计图

7.12.2　重构 Requests 请求

打开 ...\testAPI\Utils 中的 page.py 文件，编写 page.py 文件代码，示例如下：

```python
import os,requests
class Helper(object):
    def get(self,url,headers=''):
        ''' 重构 GET 请求 '''
        if url:
            r = requests.get(url=url,headers=headers)
            return r
        else:
            try:
                print('接口地址有误！')
            except Exception as M:
```

```
            print('错误原因:%s'%M)
    def post(self,url,data,headers=''):
        '''重构POST请求'''
        if url:
            r = requests.post(url=url,json=data,headers=headers)
            return r
        else:
            try:
                print('接口地址有误! ')
            except Exception as M:
                print('错误原因:%s'%M)
    def delete(self,url,headers=''):
        '''重构DELETE请求'''
        if url:
            r = requests.delete(url=url,headers=headers)
            return r
        else:
            try:
                print('接口地址有误! ')
            except Exception as M:
                print('错误原因:%s'%M)
```

对 Helper 工具类中的 Requests 常用的请求方法，如 get()、post()、delete() 请求方法进行二次封装，封装后的方法分别对应新的 get()、post()、delete() 请求方法。

7.12.3 重构接口案例

打开 ...\testAPI\TestCases 中的 test_tasks.py 文件，新增代码如下：

```python
import unittest
from testAPI.Utils.page import *   # 导入Helper工具类
class Totasks(unittest.TestCase,Helper):
    @classmethod
    def setUpClass(cls):
        pass
    @classmethod
    def tearDownClass(cls):
        pass
    def test_register(self):
        '''注册接口'''
        url = 'http://211.156.133.120:3000/register'
        data = {'username':'jackLu','password':'jackLu',
                'password_confirmation':'jackLu'}
        r = self.post(url,data)
        self.assertEqual(r.json()['username'],'jackLu')
```

```
            self.assertEqual(r.status_code,200)
    def test_login(self):
        '''登录接口'''
        url = 'http://211.156.133.120:3000/login'
        data = {'username': 'jackLu','password': 'jackLu'}
        r = self.post(url,data)
        self.assertEqual(r.status_code,200)
        self.assertEqual(r.json()['username'],'jackLu')
if __name__ == '__main__':
    unittest.main(verbosity=2)
```

Totasks 类继承 Helper 工具类。在该 Totasks 类中新增两条接口测试用例,即注册接口(test_register)和登录接口(test_login)。每条接口用例中分别使用 assertEqual() 方法对接口状态码和接口数据进行验证。

新增动态参数文件。打开 ...\testAPI\Utils 中的 page.py 文件,新增代码如下:

```
import os
class Helper(object):
    def dirname(self,fileName='',filepath='Data'):
        '''
        :param fileName: 文件名
        :param filepath: 写入指定目录
        :return:
        '''
        return os.path.join(os.path.dirname(os.path.dirname(__file__)),
                            filepath,fileName)
```

Helper 工具类中新增 dirname() 方法,其作用是将接口测试用例中服务器返回的动态数据写入指定目录下。调用 dirname() 方法时只需要传入文件名即可。

7.12.4 动态参数写入文件并读取

在接口测试过程中,有些接口是有依赖关系的,如带 Token 校验的接口。如果不登录,则无法查看登录后的所有请求信息。因为 Token 是服务器随机签发的一个"令牌"(关于 Token 的介绍具体参考 7.3 节),这个"令牌"的值每次都会变化,所以一般会将服务器生成的 Token 值单独存放在一个文件中,当其他接口用到时直接访问这个文件即可。

打开 ...\testAPI\TestCases 中的 test_tasks.py 文件,新增代码如下:

```
def writeToken(self):
    '''将 Token 写入 Token.md 文件中'''
    url = 'http://211.156.133.120:3000/login'
    data = {'username': 'jackLu','password': 'jackLu'}
    r = requests.post(url,data)
```

```python
        with open(self.dirname('Token.md'),'w') as f:
            f.write(r.json()['token'])
    def readToken(self):
        ''' 读取 Token.md 文件中的 Token 值 '''
        with open(self.dirname('Token.md'),'r') as f:
            return f.read()
```

writeToken() 方法用于将服务器返回的 Token 值写入 Token.md 文件中；调用 dirname() 方法，只需要传入要生成的文件名 Token.md 即可［文件名可以使用任意的扩展名，这里使用的是 md（markdown）语法］；readToken() 方法用于读取文件中的 Token 值。

生成的 Token.md 文件内容如下：

```
eyJhbGciOiJIUzI1NiIsInR5cCI6IkpXVCJ9.eyJ1c2VybmFtZSI6ImphY2tMdSIsImlkIjo
zOTA0LCJpYXQiOjE1NDYzOTY5MTUsImV4cCI6MTYzMjc5NjkxNX0.paXqyuEWFIAm_mI2z-
YkspnT3pjXU6s8XP4fIDkDkf8
```

7.12.5 处理接口上下游参数依赖

在整个产品业务模型中，很多接口业务是依赖上下游动态参数来实现的。在本小节接口案例中，新增创建文章接口测试用例和删除文章接口测试用例。

删除文章接口的业务逻辑是创建文章成功后，服务器会返回一个创建文章成功后的动态 ID 值，删除文章接口需要用到这个动态 ID 值（创建文章 ID），然后再删除文章。如何将 ID 值获取到并引用到当前接口中，这涉及动态参数的依赖处理。

打开 ...\testAPI\TestCases 中的 test_tasks.py 文件，新增代码如下：

```python
def test_postApiTasks(self):
    ''' 创建文章接口 '''
    url = 'http://211.156.133.120:3000/api/tasks'
    data = {'title':'https://www.cnblogs.com/fighter006','desc':' 接口描述 '}
    r = self.post(url,data,self.readToken())
    self.assertEqual(r.json()['desc'],' 接口描述 ')
    self.assertEqual(r.status_code,200)
def writeTaskId(self):
    ''' 写入 Token 到 taskID 文件中 '''
    url = 'http://211.156.133.120:3000/api/tasks'
    data = {'title': 'https://www.cnblogs.com/fighter006','desc': ' 接口描述 '}
    r = requests.post(url,data,self.readToken())
    with open(self.dirname('taskID'),'w') as f:
        f.write(str(r.json()['id']))
def getTaskID(self):
    ''' 读取 taskID '''
    with open(self.dirname('taskID'),'r') as f:
```

```
        return  f.read()
    def test_deleteApiTasks(self):
        ''' 删除文章接口 '''
        url = 'http://211.156.133.120:3000/api/tasks/:'
        r = self.delete(url+self.getTaskID(),self.readToken())
        self.assertEqual(r.status_code,200)
```

writeTaskId() 方法用于将创建后服务器返回的文章 ID 写入 taskID 文件中，getTaskID() 方法用于读取服务器返回的文章 ID。接口安全认证规定，登录后的所有请求都必须包含 Token，创建文章接口和删除文章接口必须带上 self.readToken() 方法返回的 Token 才可以访问。

7.12.6　重构 Excel 工具类

在实际接口测试过程中，接口测试用例需要用到的数据方法很多，如使用 YAML、DDT、Excel、TXT 等方法。本小节通过构造 Excel 工具类来演示接口测试过程中对数据的管理。

将接口测试数据增加到 Excel 表中，如图 7.12 所示。

请求地址	是否携带请求数据	是否携带Token	请求方式
http://211.149.163.145:3000/register	{"username":"jackLu","password":"jackLu","password_confirmation":"jackLu"}	"无"	post
http://211.149.163.145:3000/login	{"username": "jackLu", "password":"jackLu"}	"无"	post
http://211.149.163.145:3000/api/tasks	否	{"token":"asdfasdfsdds"}	get
http://211.149.163.145:3000/api/tasks	{"title":"https://www.cnblogs.co/fighter007","desc":"接口描述"}	{"token":"dsfsdsdxttru"}	post
http://211.149.163.145:3000/api/tasks/:	否	{"token":"ryrtyrtrygfng"}	delete

图 7.12　接口测试数据存放 Excel 界面

打开并编辑 ...\testAPI\Utils 中的 excles.py 文件，示例如下：

```
import json    # 导入 JSON 库
import xlrd
class Excels(object):
    ''' 构造 Excel 工具类 '''
    def readExcelData(self,rowx):
        ''' 读取 rowx 行数 '''
        book = xlrd.open_workbook(r'F:\project\testAPI\Data\data.xlsx')
        table = book.sheet_by_index(0)
        return table.row_values(rowx)
    def readUrl(self,rowx):
        ''' 读取接口地址 '''
        return self.readExcelData(rowx)[0]
    def readData(self,rowx):
        ''' 读取请求参数 '''
        return json.loads(self.readExcelData(rowx)[1])
    def readToken(self,rowx):
```

```
    ''' 读取 Token '''
    return json.loads(self.readExcelData(rowx)[2])
```

在 Excel 工具类中，readExcelData() 方法返回每一行的数据；readUrl() 方法的入参 rowx 表示第几行的第一列数据，[0] 表示第一列数据；readData() 方法用于读取第二列；json.loads() 方法将读取后的数据（字符串类型）转换为字典类型；readToken() 方法用于读取 Token 数据（图 7.12 中的 Token 值是随意写进去的，并不是真实存在的。这里是为了方便后续介绍接口测试中 Token 动态参数的赋值处理策略）。

打开 ...\testAPI\TestCases 中的 test_tasks.py 文件，对其代码进行重构，示例如下：

```python
import unittest
from testAPI.Utils.page import *    # 导入 Helper 工具类
from testAPI.Utils.excles import *  # 导入 Excels 工具类
class Totasks(unittest.TestCase,Helper,Excels):
    @classmethod
    def setUpClass(cls):
        pass
    @classmethod
    def tearDownClass(cls):
        pass
    def test_register(self):
        ''' 注册接口 '''
        r = self.post(self.readUrl(1),self.readData(1))
        self.assertEqual(r.json()['username'],'jackLu')
        self.assertEqual(r.status_code,200)
    def test_login(self):
        ''' 登录接口 '''
        r = self.post(self.readUrl(2),self.readData(2))
        self.assertEqual(r.status_code,200)
        self.assertEqual(r.json()['username'],'jackLu')
    def writeToken(self):
        ''' 将 Token 写入 Token.md 文件中 '''
        r = requests.post(self.readUrl(2),self.readData(2))
        with open(self.dirname('Token.md'),'w') as f:
            f.write(r.json()['token'])
    def readToken(self):
        ''' 读取 Token.md 文件中的 Token 值 '''
        with open(self.dirname('Token.md'),'r') as f:
            return {"Authorization":"Bearer " + f.read()}
    def test_getApiTask(self):
        ''' 获取所有文章 '''
        r = self.get(self.readUrl(3),self.readToken())
        self.assertEqual(r.status_code,200)
    def test_postApiTasks(self):
```

```python
            '''创建文章接口'''
            r = self.post(self.readUrl(4),self.readData(4),self.readToken())
            self.assertEqual(r.json()['desc'],'接口描述')
            self.assertEqual(r.status_code,200)
        def writeTaskId(self):
            '''写入Token到taskID文件中'''
            r = requests.post(self.readUrl(4),self.readData(4),self.readToken())
            with open(self.dirname('taskID'),'w') as f:
                f.write(str(r.json()['id']))
        def getTaskID(self):
            '''读取taskID'''
            with open(self.dirname('taskID'),'r') as f:
                return f.read()
        def test_deleteApiTasks(self):
            '''删除文章接口'''
            r = self.delete(self.readUrl(5) + self.getTaskID(),self.readToken())
            self.assertEqual(r.status_code,200)
if __name__ == '__main__':
    unittest.main(verbosity=2)
```

Totasks 类继承 Excels 子类，并引用 Excels 工具类中的 readUrl() 方法和 readData() 方法分别替换接口测试中用到的接口地址和请求参数。

7.12.7 动态参数赋值调用

存放在 Excel 数据中的 Token 值是固定的，可随意添加，这种操作显然不合理。针对 Excel 表中存放的固定 Token 值，继续对 test_tasks.py 代码进行重构，示例如下：

```python
...
    def test_writeToken(self):
        '''将Token写入Token.md文件中'''
        r = requests.post(self.readUrl(2),self.readData(2))
        with open(self.dirname('Token.md'),'w') as f:
            return f.write(r.json()['token'])
    def read_Token(self):
        '''读取Token.md文件中的Token值'''
        with open(self.dirname('Token.md'),'r') as f:
            return f.read()
    def setToken(self,rx):
        '''对动态参数Token赋值'''
        dinfo = self.readToken(rx)
        dinfo['token'] = self.read_Token()
        return {"Authorization":"Bearer " + dinfo['token']}
    def test_getApiTask(self):
```

```python
''' 获取所有文章 '''
r = self.get(self.readUrl(3),self.setToken(3))
self.assertEqual(r.status_code,200)
```

test_getApiTask() 方法调用 Excels 工具类下的 self.setToken() 方法，其中 (3) 表示读取 Excel 数据中的第三列数据，即 Token 数据。读取固定的 Token 值之后，将服务器生成的最新 Token 值重新赋值给 dinfo['token']，这样就实现了动态参数的赋值处理。

7.12.8　日志管理功能

在接口测试过程中，可以引用 logging 日志模块来监控接口用例在运行过程中的日志信息。logging 模块是 Python 内置的标准模块，主要用于输出运行日志，可以设置输出日志的等级、日志保存路径和日志文件回滚等。

打开 ...\testAPI\Utils 中的 page.py 文件，新增 Makelog() 方法，示例如下：

```python
import logging    # 导入 logging 模块
...
def Makelog(self,log_content):
    ''' 定义 log 日志级别 '''
    # 定义日志文件
    logFile = logging.FileHandler(self.dirname('log.txt','Log'),
                            'a',encoding='utf-8')
    # 设置 log 格式
    fmt = logging.Formatter(fmt='%(asctime)s-%(name)s-%(levelname)s-
                            %(module)s:%(message)s')
    logFile.setFormatter(fmt)
    logger1 = logging.Logger('logTest',level=logging.DEBUG)   # 定义日志
    logger1.addHandler(logFile)
    logger1.info(log_content)
    logFile.close()
...
```

Makelog() 方法的作用是设置日志级别将并日志信息写入 ...\testAPI\Log 目录下。在调用 dirname() 方法时需要传入日志文件名，目录名即可。

增加日志收集后，重构 test_tasks.py 接口测试案例，代码如下：

```python
import unittest
from testAPI.Utils.page import*          # 导入 Helper 工具类
from testAPI.Utils.excles import*        # 导入 Excels 工具类
class Totasks(unittest.TestCase,Helper,Excels):
    @classmethod
    def setUpClass(cls):
        pass
```

```python
    @classmethod
    def tearDownClass(cls):
        pass
    def test_register(self):
        '''注册接口'''
        r = self.post(self.readUrl(1),self.readData(1))
        self.assertEqual(r.json()['username'],'jackLu')
        self.Makelog('接口断言：注册接口响应数据检验 jackLu')
        self.assertEqual(r.status_code,200)
        self.Makelog('接口断言：注册接口响应状态码检验 200')
    def test_login(self):
        '''登录接口'''
        r = self.post(self.readUrl(2),self.readData(2))
        self.assertEqual(r.status_code,200)
        self.Makelog('接口断言：登录接口响应状态码检验 200')
        self.assertEqual(r.json()['username'],'jackLu')
        self.Makelog('接口断言：登录接口响应数据检验 jackLu')
    def test_writeToken(self):
        '''将 Token 写入 Token.md 文件中'''
        r = requests.post(self.readUrl(2),self.readData(2))
        with open(self.dirname('Token.md'),'w') as f:
            self.Makelog('日志跟踪：将 Token 写入 Token.md 文件中')
            return f.write(r.json()['token'])
    def read_Token(self):
        '''读取 Token.md 文件中的 Token 值'''
        with open(self.dirname('Token.md'),'r') as f:
            self.Makelog('日志跟踪：读取 Token.md 文件中的 Token')
            return  f.read()
    def setToken(self,rx):
        '''对动态参数 Token 赋值'''
        dinfo = self.readToken(rx)
        self.Makelog('动态参数处理：读取 Excel 表中的 Token 值')
        dinfo['token'] = self.read_Token()
        self.Makelog('动态参数处理：对 Token 重新赋值为最新服务器生成的 Token 值')
        return {"Authorization":"Bearer " + dinfo['token']}
    def test_getApiTask(self):
        '''获取所有文章'''
        r = self.get(self.readUrl(3),self.setToken(3))
        self.assertEqual(r.status_code,200)
        self.Makelog('接口断言：获取所有文章响应状态码检验 200')
    def test_postApiTasks(self):
        '''创建文章接口'''
        r = self.post(self.readUrl(4),self.readData(4),self.setToken(4))
        self.assertEqual(r.json()['desc'],'接口描述')
        self.Makelog('接口断言：创建文章响应数据断言[接口描述]')
```

```python
            self.assertEqual(r.status_code,200)
            self.Makelog('接口断言：创建文章响应状态码检验200')
    def writeTaskId(self):
        '''写入Token到taskID文件中'''
        r = requests.post(self.readUrl(4),self.readData(4),self.setToken(5))
        with open(self.dirname('taskID'),'w') as f:
            f.write(str(r.json()['id']))
            self.Makelog('接口业务：将创建后的文章ID写入到taskID文件中')
    def getTaskID(self):
        '''读取taskID'''
        with open(self.dirname('taskID'),'r') as f:
            self.Makelog('接口业务：读取创建文章后的文章ID')
            return f.read()
    def test_deleteApiTasks(self):
        '''删除文章接口'''
        r = self.delete(self.readUrl(5) + self.getTaskID(),self.setToken(5))
        self.assertEqual(r.status_code,200)
        self.Makelog('接口断言：删除文章响应状态码检验200')
if __name__ == '__main__':
    unittest.main(verbosity=2)
```

log.txt 文件日志信息如下：

```
2019-01-03 11:56:20,666-logTest-INFO-page:接口业务：读取创建文章后的文章ID
2019-01-03 11:56:20,682-logTest-INFO-page:动态参数处理：读取Excel表中的Token值
2019-01-03 11:56:20,682-logTest-INFO-page:日志跟踪：读取Token.md文件中的Token
2019-01-03 11:56:20,683-logTest-INFO-page:动态参数处理：对Token重新赋值为最新服务器生成的Token值
2019-01-03 11:56:20,881-logTest-INFO-page:接口断言：删除文章响应状态码检验200
2019-01-03 11:56:20,903-logTest-INFO-page:动态参数处理：读取Excel表中的Token值
2019-01-03 11:56:20,905-logTest-INFO-page:日志跟踪：读取Token.md文件中的Token
2019-01-03 11:56:20,906-logTest-INFO-page:动态参数处理：对Token重新赋值为最新服务器生成的Token值
2019-01-03 11:56:21,059-logTest-INFO-page:接口断言：获取所有文章响应状态码检验200
2019-01-03 11:56:21,668-logTest-INFO-page:接口断言：登录接口响应状态码检验200
2019-01-03 11:56:21,660-logTest-INFO-page:接口断言：登录接口响应数据检验jackLu
2019-01-03 11:56:21,686-logTest-INFO-page:动态参数处理：读取Excel表中的Token值
2019-01-03 11:56:21,689-logTest-INFO-page:日志跟踪：读取Token.md文件中的Token
2019-01-03 11:56:21,690-logTest-INFO-page:动态参数处理：对Token重新赋值为最新服务器生成的Token值
2019-01-03 11:56:21,932-logTest-INFO-page:接口断言：创建文章响应数据断言[接口描述]
2019-01-03 11:56:21,934-logTest-INFO-page:接口断言：创建文章响应状态码检验200
2019-01-03 11:56:22,496-logTest-INFO-page:接口断言：注册接口响应数据检验jackLu
2019-01-03 11:56:22,498-logTest-INFO-page:接口断言：注册接口响应状态码检验200
2019-01-03 11:56:23,054-logTest-INFO-page:日志跟踪：将Token写入Token.md文件中
```

7.12.9 配置文件功能

在接口测试过程中经常会使用到的常量,包括数据库的相关信息、接口的相关信息和邮件的相关信息等。一般使用 configparser 模块进行管理。configparser 模块的使用非常简单。

在 ...\testAPI\Config\config.ini 中新建 configer.ini 文件,并输入以下配置信息:

```
[EMAIL]
mail_host = smtp.126.com
mail_user = lurui@126.com
mail_pass = qq1234xxxx        # 126 邮箱授权码
sender = luruixxxx@126.com
receiver = luruixxxxx@126.com
subject = 接口自动化测试报告
```

读取 configer.ini 文件中的代码,示例如下:

```python
import configparser  # 导入 configparser 模块
# 创建 configParser 对象
cf = configparser.ConfigParser()
# read(filename)：读文件内容
filename = cf.read("config.ini",encoding='utf-8')
print(filename)
# sections()：得到所有的 Section,以列表形式返回
sec = cf.sections()
print(sec)
# options(section)：得到 Section 下的所有 option
opt = cf.options("EMAIL")
print(opt)
# items：得到 Section 的所有键值对
value = cf.items("EMAIL")
print(value)
# get(section,option)：得到 Section 中的 option 值,返回 string/int 类型的结果
email_host = cf.get("EMAIL","mail_host")
email_password = cf.get("EMAIL","mail_pass")
email_sender = cf.get('EMAIL','sender')
email_user = cf.get('EMAIL','mail_user')
```

输出结果如下:

```
['config.ini']
['EMAIL']
['mail_host','mail_user','mail_pass','sender','receiver','subject']
[('mail_host','smtp.126.com'),('mail_user','luruifeng@126.com'),('mail_pass',
  'qq123456689'),('sender','luruifengx@126.com'),('receiver','luruifengx@126.com'),
  ('subject','接口自动化测试报告')]
```

打开 ...\testAPI\Utils 中的 page.py 文件,新增 readConfig() 方法,示例如下:

```python
def readConfig(self):
    ''' 读取配置文件中的内容 '''
    savedata = []
    config = configparser.ConfigParser()
    config.read(self.dirname('config.ini','Config'),encoding='utf-8')
    email_host = config.get("EMAIL","mail_host")
    email_password = config.get("EMAIL","mail_pass")
    email_sender = config.get('EMAIL','sender')
    email_user = config.get('EMAIL','mail_user')
    email_receiver = config.get('EMAIL','receiver')
    email_subject = config.get('EMAIL','subject')
    savedata.append(email_host)
    savedata.append(email_password)
    savedata.append(email_sender)
    savedata.append(email_user)
    savedata.append(email_receiver)
    savedata.append(email_subject)
    return savedata
```

email_host、email_password、email_sender、email_user、email_receiver 和 email_subject 分别用于读取 config.ini 文件中的主机地址、授权码、发送者邮箱、用户邮箱、接收者邮箱和邮件主题。输出结果为一个 list 对象,示例如下:

```
['smtp.126.com','qq123456689','luruifengx@126.com','luruifeng@126.com',
 'luruifengx@126.com',' 接口自动化测试报告 ']
```

7.12.10 发送接口测试报告

打开 ...\testAPI 中的 runMain.py 文件,编写 runMain.py 文件代码,示例如下:

```python
import smtplib                                        # 邮箱服务器
from email.mime.text import MIMEText                  # 邮件模板类
import unittest
from testAPI.Config.HTMLTestRunner import HTMLTestRunner
import time,os
from email.mime.multipart import MIMEMultipart        # 邮件附件类
from email.header import Header                       # 邮件头部模板
import configparser                                   # 导入 configparser 模块
from testAPI.Utils.page import *                      # 导入 Helper 类
# 发送带邮件的函数动作
def send_mail(file_new):
    f = open(file_new,'rb')
    mail_body = f.read()
```

```python
        f.close()
        # 基本信息
        smtpserver = Helper().readConfig()[0]          # 126邮箱服务器
        pwd = Helper().readConfig()[1]                  # 126邮箱授权码
        # 定义邮件主题
        msg=MIMEMultipart()
        msg['subject'] = Header(Helper().readConfig()[-1],'utf-8')
        msg['from'] = Helper().readConfig()[2]          # 必须加，不加报错。发送者的邮箱
        msg['to'] = Helper().readConfig()[3]            # 必须加，不加报错。接收者的邮箱
        # 不加msg['from']、msg['to']报错原因，是因为发件人和收件人参数没有进行定义
        # HTML邮件正文，直接发送附件的代码片段
        body = MIMEText(mail_body,"html","utf-8")
        msg.attach(body)
        att = MIMEText(mail_body,"base64","utf-8")
        att["Content-Type"] = "application/octet-stream"
        att["Content-Disposition"] = 'attachment;filename="Interface_report.html"'
        msg.attach(att)
        # 链接邮箱服务器发送邮件
        smtp = smtplib.SMTP()
        smtp.connect(smtpserver)
        smtp.login(msg['from'],pwd)
        smtp.sendmail(msg['from'],msg['to'],msg.as_string())
        print(" 邮件发送成功 ")
# 查找最新邮件
def new_file(test_dir):
    result_dir = test_dir
    lists = os.listdir(result_dir)  # print(lists)   # 列出测试报告目录下所有的文件
    lists.sort()                                     # 从小到大排序文件
    file = [x for x in lists if x.endswith('.html')] # for循环遍历以.html格式的测试报告
    file_path = os.path.join(result_dir,file[-1])    # 找到测试报告目录下最新的测试报告
    return file_path                                 # 返回最新的测试报告
if __name__ == '__main__':
    base_dir = os.path.dirname(os.path.realpath(__file__))     # 获取文件当前路径
    test_dir = os.path.join(base_dir,'TestCases')              # 测试用例所在路径
    test_report = os.path.join(base_dir,'Reports')             # 测试报告所在路径
    testlist = unittest.defaultTestLoader.discover(test_dir,pattern = 'test*.py')
    now = time.strftime("%Y-%m-%d %H_%M_%S")
    filename = test_report + "\\" + now + 'result.html'
    fp = open(filename,'wb')
    runner = HTMLTestRunner(stream = fp,
                            title = u' 接口自动化测试框架设计报告 ',
                            description = u' 系统环境:Win10用例执行情况:')
    runner.run(testlist)
    fp.close()
    new_report = new_file(test_report)                         # 获取最新报告文件
    send_mail(new_report)                                      # 发送最新的测试报告
```

运行主文件 runMain.py，生成的测试报告如图 7.13 和图 7.14 所示。

图 7.13　接口测试报告界面

图 7.14　126 邮箱带附件的邮件界面

第8章 Robot Framework接口自动化实战

Robot Framework 中的 RequetsLibrary 库提供了多种关键字用来支持实际工作中的接口自动化测试工作。本章演示 RequestsLibrary 库在接口测试中的应用，以及接口测试数据和测试脚本的分离实战等。在学习 RequestsLibrary 库之前，首先进行 Collections 库和 ExcelLibrary 库的关键字学习。

Collections 库案例实战

Collections 库是 Robot Framework 用来处理列表和字典的库，在接口自动化测试过程中应用比较广泛。详细可参见官方介绍，该库的官方地址为 http://robotframework.org/robotframework/latest/libraries/Collections.html。

1. 导入 Collections 库

Collections 库属于 Robot Framework 内置库，在使用 Collections 库时无须安装，直接导入即可使用（注意区分大小写），如图 8.1 所示。

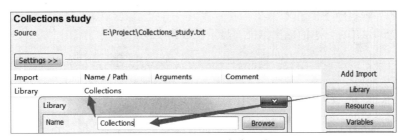

图 8.1　Collections 库导入界面

2. Collections 关键字案例

Collections 库提供的关键字非常多，本节主要演示在接口测试过程中一些常见关键字的用法及案例说明。

（1）Append To List。

Append To List 关键字用于将值添加到列表，案例如图 8.2 所示。

	Append To List			
	Settings >>			
1	${list_A}	Create List	Robot	Jmeter	Python
2	Append To List	${list_A}	Framewrok	Scripts	Selenium
3	Log	${list_A}			

图 8.2　Append To List 关键字案例

图 8.2 中，Create List 关键字用于创建一个列表 ${list_A}，通过 Append To List 关键字向 ${list_A}

追加多个元素。

按 F8 键，查看 RIDE 日志的输出结果：

```
Starting test: Project.Collections study.Append To List
20190131 19:30:50.780 : INFO : ${list_A} = [u'Robot',u'Jmeter',u'Python']
20190131 19:30:50.782 : INFO : [u'Robot',u'Jmeter',u'Python',u'Framewrok',
u'Scripts',u'Selenium']
Ending test: Project.Collections study.Append To List
```

（2）Count Values In List。

Count Values In List 关键字用于计算某一个值在列表中重复的次数，案例如图 8.3 所示。

	Count Values In List				
1	${list_A}	Create List	Robot	Jmeter	Python
2	${new_list}	Count Values In List	${list_A}	Robot	
3	Log	${new_list}			

图 8.3　Count Values In List 关键字案例

图 8.3 中，Count Values In List 关键字用于计算在 ${list_A} 列表中 Robot 字符串出现的次数（频次），并把结果赋值给新的变量 ${new_list}。

按 F8 键，查看 RIDE 日志的输出结果（显示次数为 1 次）：

```
Starting test: Project.Collections study.Count Values In List
20190131 19:38:26.758 : INFO : ${list_A} = [u'Robot',u'Jmeter',u'Python']
20190131 19:38:26.760 : INFO : ${new_list} = 1
20190131 19:38:26.761 : INFO : 1
Ending test: Project.Collections study.Count Values In List
```

（3）Get Dictionary Items。

Get Dictionary Items 关键字用于返回指定字典的项。该方法等同于 Python 字典中的 items() 方法（返回对应的键和值），案例如图 8.4 所示。

	Get Dictionary Items				
1	${dict_A}	Create Dictionary	name=fighter007	address=shenzhen	hobby=study
2	${dict_all}	Get Dictionary Items	${dict_A}		
3	Log	${dict_all}			

图 8.4　Get Dictionary Items 关键字案例

图 8.4 中，Create Dictionary 关键字用于创建字典。其中 name=fighter007 中 name 是字典中的键，fighter007 是键对应的值。

按 F8 键，查看 RIDE 日志的输出结果：

```
Starting test: Project.Collections study.Get Dictionary Items
20190131 19:48:18.141 : INFO : ${dict_A} = {u'name': u'fighter007',u'address':
u'shenzhen',u'hobby': u'study'}
20190131 19:48:18.143 : INFO : ${dict_all} = [u'address',u'shenzhen',u'hobby',
u'study',u'name',u'fighter007']
20190131 19:48:18.144 : INFO : [u'address',u'shenzhen',u'hobby',u'study',u'name',
u'fighter007']
Ending test:   Project.Collections study.Get Dictionary Items
```

（4）Get Dictionary keys。

Get Dictionary keys 关键字用于返回字典中的所有键（keys），案例如图 8.5 所示。

	Get Dictionary keys				
	Settings >>				
1	${dict_A}	Create Dictionary	name=fighter007	address=shenzhen	hobby=study
2	${dict_all}	Get Dictionary Keys	${dict_A}		
3	Log	${dict_all}			

图 8.5　Get Dictionary keys 关键字案例

图 8.5 中，Get Dictionary keys 关键字返回的结果是一个列表对象。

按 F8 键，查看 RIDE 日志的输出结果：

```
Starting test: Project.Collections study.Get Dictionary keys
20190131 19:56:53.872 : INFO : ${dict_A} = {u'name': u'fighter007',u'address':
u'shenzhen',u'hobby': u'study'}
20190131 19:56:53.873 : INFO : ${dict_all} = [u'address',u'hobby',u'name']
20190131 19:56:53.874 : INFO : [u'address',u'hobby',u'name']
Ending test:   Project.Collections study.Get Dictionary keys
```

（5）Get Dictionary Values。

Get Dictionary Values 关键字用于返回字典中所有的值（values），案例如图 8.6 所示。

	Get Dictionary Values				
	Settings >>				
1	${dict_A}	Create Dictionary	name=fighter007	address=shenzhen	hobby=study
2	${dict_all}	Get Dictionary Values	${dict_A}		
3	Log	${dict_all}			

图 8.6　Get Dictionary Values 关键字案例

按 F8 键，查看 RIDE 日志的输出结果：

```
Starting test: Project.Collections study.Get Dictionary Values
20190131 20:01:43.492 : INFO : ${dict_A} = {u'name': u'fighter007',u'address':
u'shenzhen',u'hobby': u'study'}
20190131 20:01:43.494 : INFO : ${dict_all} = [u'shenzhen',u'study',u'fighter007']
```

```
20190131 20:01:43.496 : INFO : [u'shenzhen',u'study',u'fighter007']
Ending test:   Project.Collections study.Get Dictionary Values
```

（6）Get From List。

Get From List 关键字用于返回由列表中的索引指定的值，案例如图 8.7 所示。

		Get From List			
1	${list_A}	Create List	RobotFramework	Jmeter	Python
2	${list_B}	Create List	www.cnblogs.com/fighter	IT	PO
3	${res}	Get From List	${list_B}	0	
4	Log	${res}			

图 8.7　Get From List 关键字案例

图 8.7 中，创建了 ${list_A} 和 ${list_B} 两个列表，Get From List 关键字用于访问 ${list_B} 列表中的 0 号索引位。

按 F8 键，查看 RIDE 日志的输出结果：

```
Starting test: Project.Collections study.Get From List
20190131 20:07:20.066 : INFO : ${list_A} = [u'RobotFramework',u'Jmeter',u'Python']
20190131 20:07:20.069 : INFO : ${list_B} = [u'www.cnblogs.com/fighter007',u'IT',u'PO']
20190131 20:07:20.071 : INFO : ${res} = www.cnblogs.com/fighter007
20190131 20:07:20.072 : INFO : www.cnblogs.com/fighter007
Ending test:   Project.Collections study.Get From List
```

（7）Get From Dictionary。

Get From Dictionary 关键字用于获取给定的键，并返回给定字典中的值，案例如图 8.8 所示。

		Get From Dictionary			
1	${dict_A}	Create Dictionary	name=fighter007	address=shenzhen	hobby=study
2	${res}	Get From Dictionary	${dict_A}	name	
3	Log	${res}			

图 8.8　Get From Dictionary 关键字案例

图 8.8 中，Get From Dictionary 关键字用于获取 ${dict_A} 字典中 name 键对应的值。

按 F8 键，查看 RIDE 日志的输出结果：

```
Starting test: Project.Collections study.Get From Dictionary
20190131 20:13:05.371 : INFO : ${dict_A} = {u'name': u'fighter007',u'address': u'shenzhen',u'hobby': u'study'}
20190131 20:13:05.376 : INFO : ${res} = fighter007
20190131 20:13:05.377 : INFO : fighter007
Ending test:   Project.Collections study.Get From Dictionary
```

(8) Convert To Dictionary。

Convert To Dictionary 关键字用于将给定的条目转换为 Python 字典类型，案例如图 8.9 所示。

1	${dictA}	Create Dictionary	status=0	message=successful	resultCode=1001
2	${dictB}	Create Dictionary	status=500	message=failed	resultCode=5004
3	${B}	Convert To Dictionary	${dictA}		
4	Log	${B}			

图 8.9　Convert To Dictionary 关键字案例

按 F8 键，查看 RIDE 日志的输出结果：

```
Starting test: Project.Collections study.Convert To Dictionary
20190131 20:23:08.877 :    INFO : ${dictA} = {u'status': u'0',u'message':
u'successful',u'resultCode': u'1001'}
20190131 20:23:08.880 :    INFO : ${dictB} = {u'status': u'500',u'message':
u'failed',u'resultCode': u'5004'}
20190131 20:23:08.881 :    INFO : ${B} = {u'status': u'0',u'message':
u'successful',u'resultCode': u'1001'}
20190131 20:23:08.882 :    INFO : {u'status': u'0',u'message': u'successful',
u'resultCode': u'1001'}
Ending test:  Project.Collections study.Convert To Dictionary
```

8.2　ExcelLibrary 库案例应用

Robot Framework 中的 ExcelLibrary 是一个操作 Excel 表的库，它提供关键字，允许打开、读取、写入和保存 Excel 文件。

1. 安装 ExcelLibrary 库

ExcelLibrary 库可以直接通过 pip 命令来安装。打开 cmd 命令提示符界面，输入"pip install robotframework-ExcelLibrary"进行在线安装，示例如下：

```
C:\Users\Administrator>pip install robotframework-ExcelLibrary
Collecting robotframework-ExcelLibrary
  Using cached https://files.pythonhosted.org/packages/b8/e7/8c079a814e7ad288ec2
fc15671d8dc526e3d537bb00e4ab2b209a63674ed/robotframework-excellibrary-0.0.2.zip
Requirement already satisfied: robotframework>=2.8.5 in c:\python27\lib\site-pac
kages (from robotframework-ExcelLibrary)
Collecting xlutils>=1.7.1 (from robotframework-ExcelLibrary)
```

```
Using cached https://files.pythonhosted.org/packages/c7/55/e22ac73dbb316cabb5d
100% |████████████████████████████████| 102kB
123kB/s
Installing collected packages: xlrd,xlwt,xlutils,natsort,robotframework-
ExcelLibrary
  Running setup.py install for robotframework-ExcelLibrary ... done
Successfully installed natsort-5.5.0 robotframework-ExcelLibrary-0.0.2 xlrd-1.2.0
xlutils-2.0.0 xlwt-1.3.0
```

2. 导入 ExcelLibrary 库

单击"Library"按钮,在"Name"文本框中输入"ExcelLibrary"(注意区分大小写),如图 8.10 所示。

图 8.10　导入 ExcelLibrary 库界面

3. ExcelLibrary 关键字案例

(1) Open Excel。

Open Excel 关键字用于打开本地指定目录下的 Excel 文件,案例如图 8.11 所示。

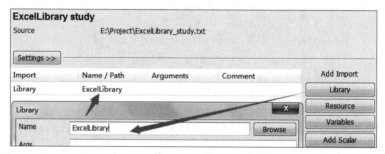

图 8.11　Open Excel 关键字案例

图 8.11 中,Open Excel Current Directory 关键字只能读取测试项目所在目录下的 Excel 文件。如果读取的 Excel 文件不和测试项目处于同级目录,则打开报错。

按 F8 键,查看 RIDE 日志的输出结果:

```
Starting test: Project.ExcelLibrary study.Open Excel
20190201 09:46:09.355 : INFO : Opening file at E:\Project\ExecelData.xls
Ending test:  Project.ExcelLibrary study.Open Excel
```

（2）Get Row Count。

Get Row Count 关键字用于返回指定工作表名称的特定行数，案例如图 8.12 所示。

Get Row Count

Settings >>

1	#打开ExcelData.xls		
2	Open Excel	E:\\Project\\ExecelData.xls	
3	#获取指定sheet行数		
4	${cout_num}	Get Row Count	Sheet1
5	Log	${cout_num}	

图 8.12　Get Row Count 关键字案例

图 8.12 中，Get Row Count 关键字用于读取 ExcelData.xls 文件中的第一个 Sheet1 对象，在获取所有行数后返回。

按 F8 键，查看 RIDE 日志的输出结果：

```
Starting test: Project.ExcelLibrary study.Get Row Count
20190201 09:52:44.867 :  INFO : ${cout_num} = 11
20190201 09:52:44.867 :  INFO : 11
Ending test:  Project.ExcelLibrary study.Get Row Count
```

（3）Get Column Count。

Get Column Count 关键字用于返回指定工作表名称的特定列数，案例如图 8.13 所示。

Get Column Count

Settings >>

1	#打开ExcelData.xls		
2	Open Excel	E:\\Project\\ExecelData.xls	
3	#获取指定sheet列数		
4	${column_num}	Get Column Count	Sheet1
5	Log	${column_num}	

图 8.13　Get Column Count 关键字案例

按 F8 键，查看 RIDE 日志的输出结果：

```
Starting test: Project.ExcelLibrary study.Get Column Count
20190201 09:57:00.785 :  INFO : ${column_num} = 7
20190201 09:57:00.785 :  INFO : 7
Ending test:  Project.ExcelLibrary study.Get Column Count
```

（4）Get Row Values。

Get Row Values 关键字用于获取并返回指定工作表名称的特定行值，案例如图 8.14 所示。

Get Row Values

Settings >>

1	#打开ExcelData.xls			
2	Open Excel	E:\\Project\\ExecelData.xls		
3	#获取指定行的数据			
4	${row_zero}	Get Row Values	Sheet1	1
5	Log	${row_zero}		

图 8.14　Get Row Values 关键字案例

图8.14 中，Get Row Values 关键字用于读取 ExcelData.xls 文件下的 Sheet1 对象中的第二行数据。1 表示索引位，实际是读取第二行数据。

按 F8 键，查看 RIDE 日志的输出结果：

```
Starting test: Project.ExcelLibrary study.Get Row Values
20190201 10:00:55.554 :  INFO : ${row_zero} = [('A2',u'1'),('B2',u'http://192.168.0.
108:8000'),('C2',u'/login'),('D2',u'zhangshan'),('E2',u'"status":0'),('F2',u'"message":
"successful"'),('G2',u'"code":0')]
20190201 10:00:55.554 :  INFO : [('A2',u'1'),('B2',u'http://192.168.0.108:8000'),
('C2',u'/login'),('D2',u'zhangshan'),('E2',u'"status":0'),('F2',u'"message":
"successful"'),('G2',u'"code":0')]
Ending test:  Project.ExcelLibrary study.Get Row Values
```

（5）Get Column Values。

Get Colunm Values 关键字用于读取某一列数据，案例如图 8.15 所示。

Get Coulmn Values

Settings >>

1	#打开ExcelData.xls			
2	Open Excel	E:\\Project\\ExecelData.xls		
3	#获取指定列中的数据			
4	${row_zero}	Get Column Values	Sheet1	1

图 8.15　Get Column Values 关键字案例

图 8.15 中，Get Column Values 关键字用于读取 Sheet1 对象中的第二列数据。

按 F8 键，查看 RIDE 日志的输出结果：

```
Starting test: Project.ExcelLibrary study.Get Coulmn Values
2019020110:08:21.382:INFO:${row_zero} = [('B1',u'\u57df\u540d'),('B2',u'http://192.
168.0.108:8000'),('B3',u'http://192.168.0.108:8001'),('B4',u'http://192.168.0.108:
8002'),('B5',u'http://192.168.0.108:8003'),('B6',u'http://192.16...
20190201 10:08:21.382 :  INFO : [('B1',u'\u57df\u540d'),('B2',u'http://192.168.
0.108:8000'),('B3',u'http://192.168.0.108:8001'),('B4',u'http://192.168.0.108:
8002'),('B5',u'http://192.168.0.108:8003'),('B6',u'http://192.168.0.108:8004'),
('B7',u'http://192.168.0.108:8005'),('B8',u'http://192.168.0.108:8006'),('B9',
u'http://192.168.0.108:8007'),('B10',u'http://192.168.0.108:8008'),('B11',
```

```
u'http://192.168.0.108:8009')]
Ending test:   Project.ExcelLibrary study.Get Coulmn Values
```

（6）Read Cell Data By Coordinates。

Read Cell Data By Coordinates 关键字通过指定列、行编号来获取指定单元格的值，案例如图 8.16 所示。

Read Cell Data By Coordinates

Settings >>

1	#打开ExcelData.xls				
2	Open Excel	E:\\Project\\ExecelData.xls			
3	#读取指定单元格数据				
4	${res}	Read Cell Data By Coordinates	Sheet1	1	1
5	Log	${res}			

图 8.16　Read Cell Data By Coordinates 关键字案例

图 8.16 中，Read Cell Data By Coordinates 关键字有 3 个参数，其中 Sheet1 表示打开的指定 Sheet 对象，第一个 1 表示读取的行数是第二行，第二个 1 表示读取的列数是第二列。

按 F8 键，查看 RIDE 日志的输出结果：

```
Starting test: Project.ExcelLibrary study.Read Cell Data By Coordinates
20190201 10:13:00.873 :  INFO : ${res} = http://192.168.0.108:8000
20190201 10:13:00.873 :  INFO : http://192.168.0.108:8000
Ending test:   Project.ExcelLibrary study.Read Cell Data By Coordinates
```

8.3 RequestsLibrary 库案例实战

Robot Framework 中的 RequestsLibrary 库可以用来进行接口测试，使用起来也非常简单。在使用 RequestsLibrary 库前需要先安装相应的 RequetsLibrary 库。

1. 安装 RequestsLibrary 库和 Requets 库

推荐使用在线方式安装。打开 cmd 命令提示符界面，输入“pip install robotframework-requests”进行在线安装，示例如下：

```
C:\Users\Administrator>pip install robotframework-requests
Collecting robotframework-requests
  Using cached https://files.pythonhosted.org/packages/ca/72/cc94e0de0bc9d38d33f
7133a233089d9b1be17345d609af2bb54a3fff932/robotframework-requests-0.5.0.tar.gz
```

```
Requirement already satisfied: robotframework in c:\python27\lib\site-packages
(from robotframework-requests)
Requirement already satisfied: requests in c:\python27\lib\site-packages (from
robotframework-requests)
Requirement already satisfied: certifi>=2017.4.17 in c:\python27\lib\site-packages
(from requests->robotframework-requests)
Installing collected packages: robotframework-requests
  Running setup.py install for robotframework-requests ... done
Successfully installed robotframework-requests-0.5.0
```

安装 Requests 库，继续使用 pip 命令安装 pip install requests，示例如下：

```
C:\Users\Administrator>pip install requests
Collecting requests
  Using cached https://files.pythonhosted.org/packages/7d/e3/20f3d364d6c8e5d2353
c72a67778eb189176f08e873c9900e10c0287b84b/requests-2.21.0-py2.py3-none-any.whl
Requirement already satisfied: urllib3<1.25,>=1.21.1 in c:\python27\lib\site-
packages (from requests)
Requirement already satisfied: idna<2.9,>=2.5 in c:\python27\lib\site-packages
(from requests)
Installing collected packages: requests
Successfully installed requests-2.21.0
```

2. 导入 RequetsLibrary 库和 Requests 库

单击 "Library" 按钮，在 "Name" 文本框中输入相应关键字名称，分别导入 RequetsLibrary 库和 Requests 库（注意区分大小写），如图 8.17 所示。

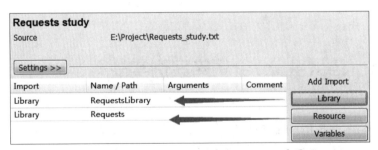

图 8.17　导入 RequestLibrary 库和 Requests 库界面

3. 发送 GET 请求案例

在实际接口测试过程中，GET 请求可以带参数传递，也可以不带参数传递。本案例演示使用 requests.Get 关键字发送一个带参数的 GET 请求，案例如图 8.18 所示。

	发送GET请求		
	Settings >>		
1	#使用Collections设置参数		
2	${params}	Create Dictionary	wd=python
3	#使用Requets库中的get关键字发送请求		
4	${result}	requests.Get	http://www.baidu.com/s? params=${params}
5	#返回的文本字符串格式		
6	Log	${result.text}	
7	#校验返回的结果是否正常		
8	Should Be Equal As Numbers	${result.status_code}	200

图 8.18 GET 请求案例

图 8.18 中，Create Dictionary 关键字用于创建一个字典 ${parms} 并将其作为请求参数传入 params 中，requests.Get 关键字发送接口需要两个参数（一个是接口地址，另一个是接口参数），{result.text} 表示把服务器的响应报文按照文本字符串的方式返回，最后通过 Should Be Equal As Numbers 关键字来断言响应的状态码是否是 200。

4. 发送 POST 请求案例

POST 请求都是带参数进行传递的，并且将参数存储在请求体中传递。在 Robot Framework 中，可以使用 request.Post 关键字来发送 POST 请求，案例如图 8.19 所示。

	发送POST请求			
	Settings >>			
1	#使用Collections设置参			
2	${params}	Create Dictionary	dateUpdated=2019-2-1	
3	#发送POST请求			
4	${result}	requests.Post	http://www.pingan.com/cm...opList.do	params=${params}
5	#将结果转换为JSON格			
6	${res}	To Json	${result.content}	
7	Log	${res}		
8	#接口断言			
9	Should Be Equal As Numbers	${result.status_code}	200	
10	Should Contain	${result.content}	"resultCode":"0"	

图 8.19 POST 请求案例

图 8.19 中，将 ${params} 作为参数传入 ${result} 接口中，然后使用 To Json 关键字将接口响应信息转换为 JSON 格式，最后使用 Should Be Equals As Numbers 和 Should Contain 关键字对响应状态码和响应字段断言。

5. 发送带 Headers 接口案例

有些接口带 Headers 信息，在 Robot Framework 中发送带请求头的接口，案例如图 8.20 所示。

![带Headers的POST请求示例表格]

图 8.20 带 Headers 请求头接口案例

图 8.20 中，使用 Create Dictionary 关键字组装信息头字段 ${headers}，${body} 为请求参数，使用 JSON 解码器将 ${body} 作为请求参数传入 ${result} 中，发送该接口必须带 Headers 及 ${body} 信息。最后对接口进行断言。${result.json()}["ResultCode"] 表示获取字典中的"ResultCode"键对应的值是否为数字。

ExcelLibrary 库数据管理案例实战

在接口测试过程中，随着接口案例的不断增多，接口测试数据的维护越来越耗费精力。通过前面的学习，可知可以使用 Robot Framework 框架提供的 ExcelLibrary 库中的关键字方法来操作测试数据文件，从而实现接口测试数据与接口脚本的分离。

（1）准备好接口数据文档。

将接口地址、请求参数和预期值写入 Excel 文件中，如图 8.21 所示。

用例编号	接口地址	请求参数	预期值
注册接口	http://211.149.123.185:3000/register	jackLu	"username":"jackLu"
登录接口	http://211.149.123.185:3000/login	jackLu	"username":"jackLu"
更新商户接口	http://www.dateUpdate.com/cms-tmplt/pinganlife/synShopList.do	2019-2-1	"resultCode":"0"

图 8.21 接口测试数据界面

在使用 ExcelLibrary 库时需要注意以下几点。

①目前只支持 .xls 格式的 Excel 文档。

②Excel 表格中的数值需设置单元格格式为文本，否则代码读取后会显示为浮点型，如 100,

读取后显示为 100.00，造成传参不一致。

③读取每行或每列数据，存储到 list 变量中，一般为二维数组，如 ${result} = [('A2', u'1'),('B2', u'http://192.168.1.46：2000'), ('C2', u'/doctor_search')]。

④数组的位置编号都从 0 开始。

（2）相关测试库的导入。

将 ExcelLibrary 库及相关库，如 RequestsLibrary、Requests 和 Collections 导入测试套件中，如图 8.22 所示。

图 8.22　导入相关库界面

（3）封装 Excel 数据关键字。

创建资源文件"common.txt"，在该资源文件下新建"登录：用户注册"用户关键字，并填入接口测试代码，案例如图 8.23 所示。

图 8.23　登录：用户注册接口案例

Arguments 选项中的 ${Excelpath}、${sheetName} 和 ${rowx} 分别表示 Excel 文件所在的位置、读取的 Sheet 对象名称和第几行数据。

- ${row_line[1][1]}：表示读取返回结果 ${row_line} 数组中的第一个元组里的第一个元素，即接口参数 ${row_line} 可以通过运行结果查看。
- ${row_line[3][1]}：表示读取 Excel 文件中的接口对应的预期结果。

继续新增"登录 - 用户登录"用户关键字，并填入接口测试代码，案例如图 8.24 所示。

登录：用户登录

1	#打开Excel文件			
2	Open Excel	${Excelpath}		
3	#读取某一行数据			
4	${row_line}	Get Row Values	${sheetName}	${rowx}
5	#设置参数			
6	${data}	Create Dictionary	username=${row_line[2][1]}	password=${row_line[2][1]}
7	#发送POST请求			
8	${result}	requests.Post	${row_line[1][1]}	json=${data}
9	#转换JSON格式			
10	${res}	To Json	${result.content}	
11	#断言			
12	Should Be Equal As Numbers	${result.status_code}	200	
13	Should Contain	${result.content}	${row_line[3][1]}	

图 8.24　登录：用户登录接口案例

登录接口只需要传入 username 和 password 参数即可。在调用该用户关键字时，需要传入 Excel 文件所在路径、获取 Sheet 对象，以及读取的行数（注意，登录接口地址在第二行）。

继续新增"登录：更新优惠商户接口"用户关键字，并填入接口测试代码，案例如图 8.25 所示。

登录：更新优惠商户接口

1	#打开Excel文件			
2	Open Excel	${Excelpath}		
3	#读取某一行数据			
4	${row_line}	Get Row Values	${sheetName}	${rowx}
5	#设置参数			
6	${data}	Create Dictionary	dateUpdated=${row_line[2][1]}	
7	#发送POST请求			
8	${result}	requests.Post	${row_line[1][1]}	params=${data}
9	#转换JSON格式			
10	${res}	To Json	${result.content}	
11	#断言			
12	Should Be Equal As Numbers	${result.status_code}	200	
13	Should Contain	${result.content}	${row_line[3][1]}	

图 8.25　登录：更新优惠商户接口案例

（4）调用封装关键字。

首先需要引入资源文件，在 Interface_test 测试套件下导入"common.txt"资源文件，然后新增 3 条接口测试用例，如图 8.26 所示。

图 8.26　接口测试案例

3 条接口案例是注册接口、登录接口和查看优惠商户接口，分别调用相应的用户关键字并传入所需参数。

（5）行接口案例。

注册接口案例，运行结果如下：

```
Starting test: Project.Interface Test.注册接口
20190201 22:31:42.229 :   INFO : ${row_line} = [('A2',u'\u6ce8\u518c\u63a5\u53e3'),
('B2',u'http://211.149.123.185:3000/register'),('C2',u'jackLu'),('D2',u'"username":
"jackLu"')]
20190201 22:31:42.229 :   INFO : ${data} = {u'username': u'jackLu',u'password':
u'jackLu',u'password_confirmation': u'jackLu'}
20190201 22:31:42.869 :   INFO : ${result} = <Response [200]>
20190201 22:31:42.885 :   INFO : To JSON using : content={"id":12484,"username":
"jackLu","password":"$2a$10$RXHlwWHj.tnQYcyP0.PTTewjAFjdceMdamno2AQonQhZ.8Oq3BeXa",
"updatedAt":"2019-02-01T14:29:29.000Z","createdAt":"2019-02-01T14:29:29.000Z"}
20190201 22:31:42.885 : INFO : To JSON using : pretty_print=False
20190201 22:31:42.885 :   INFO : ${res} = {u'username': u'jackLu',u'password':
u'$2a$10$RXHlwWHj.tnQYcyP0.PTTewjAFjdceMdamno2AQonQhZ.8Oq3BeXa',u'id':
12484,u'createdAt': u'2019-02-01T14:29:29.000Z',u'updatedAt': u'2019-02-
01T14:29:29.000Z...
20190201 22:31:42.885 :   INFO :
Argument types are:
<type 'int'>
<type 'unicode'>
Ending test:   Project.Interface Test.注册接口
```

登录接口案例，运行结果如下：

```
Starting test: Project.Interface Test.登录接口
20190201 22:31:42.900 :    INFO : ${row_line} = [('A3',u'\u767b\u9646\u63a5\u53e3'),
('B3',u'http://211.149.123.185:3000/login'),('C3',u'jackLu'),('D3',u'"username":
"jackLu"')]
20190201 22:31:42.916 :    INFO : ${data} = {u'username': u'jackLu',u'password':
u'jackLu'}
20190201 22:31:43.510 :    INFO : ${result} = <Response [200]>
20190201 22:31:43.510 :    INFO : To JSON using : content={"id":12471,"username":
"jackLu","token":"eyJhbGciOiJIUzI1NiIsInR5cCI6IkpXVCJ9.eyJ1c2VybmFtZSI6I
mphY2tMdSIsImlkIjoxMjQ3MSwiaWF0IjoxNTQ5MDMxMzcwLCJleHAiOjE2MzU0MzEzNzB9.
qcxiYpOo6o4qS7cPoyhXzc8UVOzh-pIA8yN_QvF8CNs"}
20190201 22:31:43.510 :    INFO : To JSON using : pretty_print=False
20190201 22:31:43.510 :    INFO : ${res} = {u'username': u'jackLu',u'token': u'eyJhbG
ciOiJIUzI1NiIsInR5cCI6IkpXVCJ9.eyJ1c2VybmFtZSI6ImphY2tMdSIsImlkIjoxMjQ3MSwiaWF0Ijo
xNTQ5MDMxMzcwLCJleHAiOjE2MzU0MzEzNzB9.qcxiYpOo6o4qS7cPoyhXzc8UVOzh-pIA8yN_Q...
20190201 22:31:43.510 :    INFO :
Argument types are:
<type 'int'>
<type 'unicode'>
Ending test:   Project.Interface Test.登录接口
```

查询优惠商户接口案例，运行结果如下：

```
Starting test: Project.Interface Test.查看优惠商户接口
```

第9章
Appium自动化测试实战

本章介绍移动端自动化测试框架——Appium，通过本章的学习，读者可以具备独立搭建移动端自动化测试环境的能力，以及掌握 Appium 中常见的 WebdriverApi 应用等。本章最后演示如何使用 Appium 测试框架设计一个完整的移动端自动化测试项目。

Appium 是一个开源的、跨平台的测试框架，可以用来测试原生及混合的移动端应用。Appium 支持 iOS、Android 及 FirefoxOS 平台。Appium 支持 Selenium WebDriver 的所有语言，如 Java、Object-C、JavaScript、PHP、Python、Ruby 和 C# 等，同样也可以使用 Selenium WebDriver 的 API。如果读者使用过 Selenium，那么学习使用 Appium 时会非常容易。

9.1 安装 Appium 环境

Appium 是一个开源工具，可用于在 iOS 和 Android 平台上自动化移动本机、移动 Web 和移动混合应用程序。移动原生应用是使用 iOS 或 Android SDK 编写、使用移动浏览器访问的网络应用。移动混合应用程序具有围绕 Webview 的本机包装，一种支持与 Web 内容交互的本机控件。例如，Phonegap 等项目可以轻松地使用 Web 技术构建应用程序，然后捆绑到本机包装器中，这些是混合应用程序。重要的是，Appium 是跨平台的，允许使用相同的 API 针对多个平台（iOS、Android）编写测试，这可以在 iOS 和 Android 测试套件之间实现大量或全部的代码重用。

Appium 环境的安装相对 Selenium 环境的安装要复杂一些。Appium 可以同时支持 Android 和 iOS 端的自动化测试，但是 iOS 端自动化测试无法在 Windows 下运行，所以本节是基于 Win10 64 位操作系统下的搭建教程。

9.1.1 安装 Node.js 环境

Node.js 是一个基于 Chrome JavaScript 运行时建立的平台，因为 Appium 使用 Node.js 实现，所以安装 Appium 前必须要先保证 Node.js 环境是正常的。

Node.js 的下载地址为 https://nodejs.org/en/download，安装时要注意区分操作系统的版本，如图 9.1 所示。

本小节选择 Window Installer(.msi) 64-bit 版本进行安装，安装过程中选择默认设置即可。安装完成后，验证 Node.js，在 cmd 命令提示符界面输入"npm"，如果返回如下信息，则表示安装成功；反之，则需要将 Node.js 安装路径追加到系统环境变量中，再次验证。

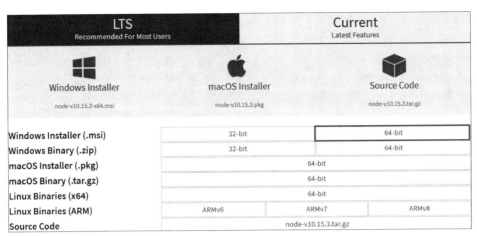

图 9.1　Node.js 下载界面

```
C:\Users\23939>npm
Usage: npm <command>
where <command> is one of:
    access,adduser,bin,bugs,c,cache,completion,config,
    ddp,dedupe,deprecate,dist-tag,docs,doctor,edit,
    explore,get,help,help-search,i,init,install,
    install-test,it,link,list,ln,login,logout,ls,
    outdated,owner,pack,ping,prefix,prune,publish,rb,
    rebuild,repo,restart,root,run,run-script,s,se,
    search,set,shrinkwrap,star,stars,start,stop,t,team,
    test,tst,un,uninstall,unpublish,unstar,up,update,v,
    version,view,whoami

npm <cmd> -h     quick help on <cmd>
npm -l           display full usage info
npm help <term>  search for help on <term>
npm help npm     involved overview

Specify configs in the ini-formatted file:
    C:\Users\23939\.npmrc
or on the command line via: npm <command> --key value
Config info can be viewed via: npm help config
npm@4.1.2 C:\Program Files\nodejs\node_modules\npm
```

注意：npm 是一个 node 包管理工具，有了 npm，可以很快地找到特定服务要使用的包并下载、安装及管理已经安装的包。

9.1.2　安装 Appium 工具

Appium 的安装方式可以选择在线安装和离线安装，笔者更推荐后者，因为在线安装要依赖网络，下载时间比较久。如果使用在线安装，直接在 cmd 命令提示符界面输入"npm -g appium"即可。

离线安装的下载地址为 https://bitbucket.org/appium/appium.app/downloads，下载界面如图 9.2 所示。

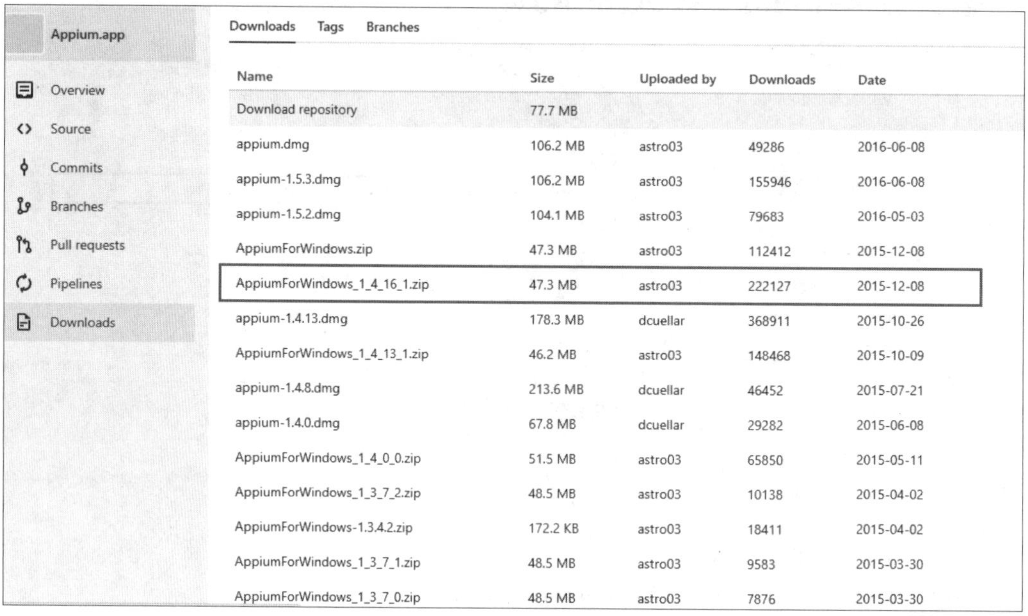

图 9.2　Appium 下载界面

Appium 当前最新版本为 AppiumForWindows_1_4_16_1.zip，这是 Windows 版本，如果是 Mac，则安装 appium-1.5.2.dmg 版本。下载完后解压 .zip 包，直接双击安装 Appium 即可。打开 Appium，首页界面如图 9.3 所示。

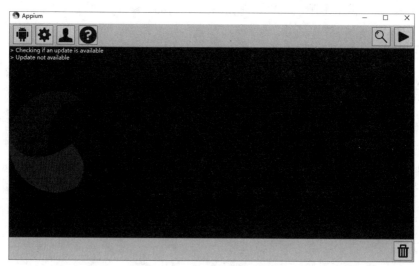

图 9.3　Appium 首页界面

安装好 Appium 工具后，需要找到 .bin 文件的安装目录 C:\Program Files (x86)\Appium\node_modules（笔者本机默认安装位置），如图 9.4 所示。

如图 9.4 所示，需要将 C:\Program Files (x86)\Appium\node_modules 变量值追加到系统的环境

变量 Path 中。

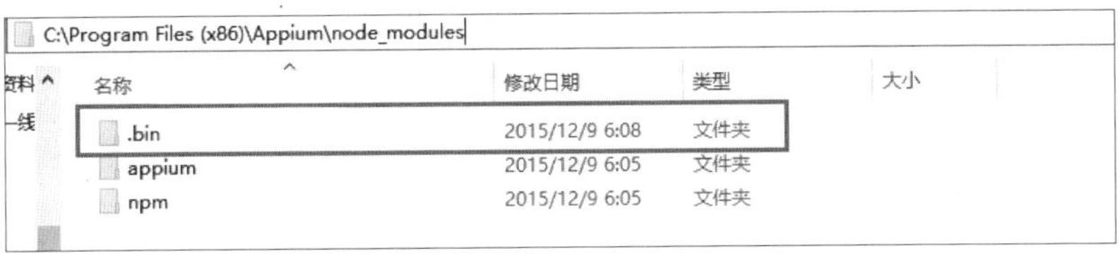

图 9.4 .bin 文件所在目录界面

9.1.3 安装 Java 环境

因为 Android 操作系统由 Java 语言开发，所以必须要安装 Java 环境。Java 的下载地址为 https://www.oracle.com/technetwork/java/javase/downloads/jdk8-downloads-2133151.html，笔者选择的 JDK 版本为 1.8，不建议选择 1.8 以下的版本。Java 安装界面如图 9.5 所示。

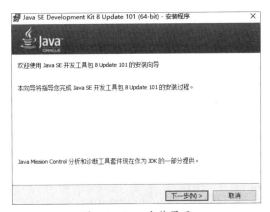

图 9.5 Java 安装界面

安装 Java 后，右击"我的电脑"在弹出的快捷菜单中选择"属性"选项，打开"系统"窗口，单击"高级系统设置"超链接，弹出"系统属性"对话框，单击"环境变量"按钮，在弹出的"环境变量"对话框中进行设置。需设置 3 项属性，分别是 JAVA_HOME、CLASSPATH 和 Path（不区分大小写），若已存在则单击"编辑"按钮；若不存在则单击"新建"按钮。新增 JAVA_HOME 变量，如图 9.6 所示。

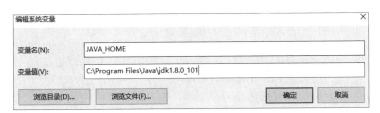

图 9.6 JAVA_HOME 设置界面

变量值 C:\Program Files\Java\jdk1.8.0_101 表示 JDK 的安装路径。在系统变量中，继续新增 CALSSPATH 变量，如图 9.7 所示。

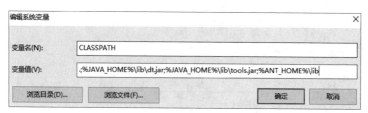

图 9.7　CLASSPATH 设置界面

在系统变量原有的 Path 变量中，追加变量值 ";%JAVA_HOME%\bin;%JAVA_HOME%\jre\bin;"，如图 9.8 所示。

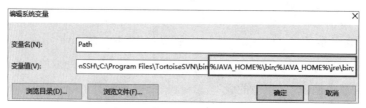

图 9.8　追加 Path 变量值界面

验证 Java 是否安装成功，在 cmd 命令提示符界面输入 "java"，示例如下：

```
C:\Users\23939>java
用法: java [-options] class [args...]
          (执行类)
    或  java [-options] -jar jarfile [args...]
          (执行 jar 文件)
其中，选项包括:
    -d32          使用 32 位数据模型 (如果可用)
    -d64          使用 64 位数据模型 (如果可用)
    -server       选择 "server" VM
                  默认 VM 是 server.
    -cp <目录和 zip/jar 文件的类搜索路径>
    -classpath <目录和 zip/jar 文件的类搜索路径>
                  用 ; 分隔的目录,JAR 档案
                  和 ZIP 档案列表,用于搜索类文件。
                  未来发行版中删除。
                  需要指定的版本才能运行
    -showversion  输出产品版本并继续
    -jre-restrict-search | -no-jre-restrict-search
    -agentpath:<pathname>[=<选项>]
                  按完整路径名加载本机代理库
    -javaagent:<jarpath>[=<选项>]
                  加载 Java 编程语言代理,请参阅 java.lang.instrument
    -splash:<imagepath>
                  使用指定的图像显示启动屏幕
```

再次输入 "javac"，示例如下：

```
C:\Users\23939>javac
用法: javac <options> <source files>
其中, 可能的选项包括:
  -s <目录>                    指定放置生成的源文件的位置
  -h <目录>                    指定放置生成的本机标头文件的位置
  -implicit:{none,class}       指定是否为隐式引用文件生成类文件
  -encoding <编码>             指定源文件使用的字符编码
  -source <发行版>             提供与指定发行版的源兼容性
  -target <发行版>             生成特定 VM 版本的类文件
  -profile <配置文件>          请确保使用的 API 在指定的配置文件中可用
  -version                     版本信息
  -help                        输出标准选项的提要
  -A 关键字[=值]               传递给注释处理程序的选项
  -X                           输出非标准选项的提要
  -J<标记>                     直接将 <标记> 传递给运行时系统
  -Werror                      出现警告时终止编译
  @<文件名>                    从文件读取选项和文件名
```

知识拓展：JDK 是 Java 开发工具包，是程序员使用 Java 语言编写 Java 程序所需的开发工具包，是提供给程序员使用的。JDK 包含了 JRE，同时还包含了编译 Java 源码的编译器 javac；还包含了很多 Java 程序调试和分析的工具，如 jconsole、jvisualvm 等；还包含了 Java 程序编写所需的文档和 demo 例子程序。如果需要运行 Java 程序，只需安装 JRE 即可；如果需要编写 Java 程序，需要安装 JDK。

9.1.4　安装 Android SDK

Android SDK（Software Development Kit，软件开发工具包）是为特定的软件包、软件框架、硬件平台和操作系统等建立应用软件的开发工具的集合。它提供了 Android API 库和开发工具构建、测试和调试应用程序。简单来讲，Android SDK 可以看作用于开发和运行 Android 应用的一个软件。

SDK 的下载地址为 http://www.androiddevtools.cn，下载后解压，如图 9.9 所示。

图 9.9　Android SDK 目录界面

接下来设置 Android 环境变量，设置方法与 Java 环境变量类似。本机目录为 D:\android\android-sdk-windows，右击"我的电脑"在弹出的快捷菜单中选择"属性"选项，打开"系统"窗口，单击"高级系统设置"超链接，弹出"系统属性"对话框，单击"环境变量"按钮，弹出"环境变量"对话框，单击"新建"按钮，在系统变量中新增 ANDROID_HOME 变量，如图 9.10 所示。

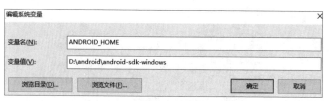

图 9.10 ANDROID_HOME 设置界面

在系统变量原有的 Path 变量中，追加变量值"; %ANDROID_HOME%\platform-tools;%ANDROID_HOME%\tools;" 如图 9.11 所示。

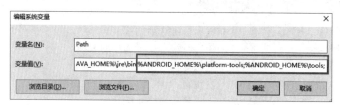

图 9.11 设置 Android 路径到 Path 界面

9.1.5 安装 Android SDK Platform-Tools

Android SDK Platform-Tools 包含 adb、fastboot 等工具包。把解压出来的 platform-tools 文件夹放在 Android SDK 根目录下，并把 adb 所在的目录添加到系统 Path 路径中，即可在 cmd 命令提示符界面直接访问 adb、fastboot 等工具，如图 9.12 所示。

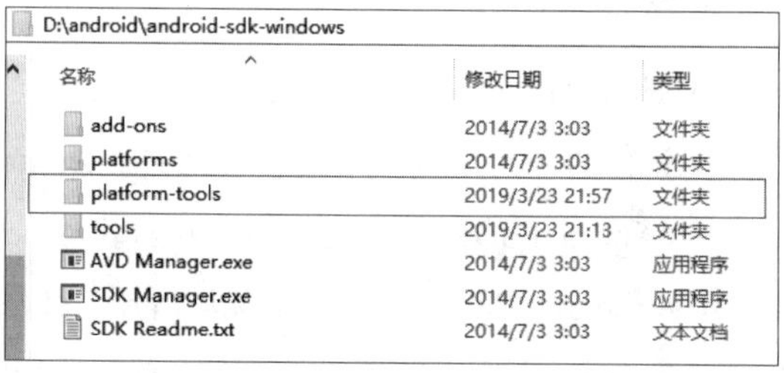

图 9.12 paltfrom-tools 设置界面

paltfrom-tools 的下载地址为 https://www.androiddevtools.cn，应根据操作系统的相应型号选择对应的 platform-tools 版本。笔者下载的是 platform-tools_r20-windows.zip 版本。最后，打开 cmd 命令提示符界面，输入 "appium-doctor"，检查 Appium 环境，示例如下：

```
C:\Users\23939>appium-doctor
Running Android Checks
√ ANDROID_HOME is set to "D:\android\android-sdk-windows"
√ JAVA_HOME is set to "C:\Program Files\Java\jdk1.8.0_101."
√ ADB exists at D:\android\android-sdk-windows\platform-tools\adb.exe
√ Android exists at D:\android\android-sdk-windows\tools\android.bat
```

```
√ Emulator exists at D:\android\android-sdk-windows\tools\emulator.exe
√ Android Checks were successful.
√ All Checks were successful
```

All Checks were succerssful 表示 Appium 自动化测试环境安装成功。

9.1.6 安装 Appium 工具 Client

本小节是基于 Python 语言驱动的,所以应确保计算机具备 Python 环境(建议使用最新的 Python 3 环境)。安装 Python 环境,这里推荐在线安装。打开 cmd 命令提示符界面,输入"pip install Appium-Python-Client"进行在线安装,示例如下:

```
C:\Users\23939>pip install Appium-Python-Client
Collecting Appium-Python-Client
  Downloading https://files.pythonhosted.org/packages/26/f1/f932791ec73be6e13539fb20
1f6923305b8e67b2b47078fd2efc3ad4f865/Appium-Python-Client-0.40.tar.gz (41kB)
    100% |████████████████████████████████| 51kB 107kB/s
Requirement already satisfied: selenium>=3.14.1 in c:\python36\lib\site-packages
(from Appium-Python-Client) (3.141.0)
Requirement already satisfied: urllib3 in c:\python36\lib\site-packages
(from selenium>=3.14.1->Appium-Python-Client) (1.23)
Installing collected packages: Appium-Python-Client
  Running setup.py install for Appium-Python-Client ... done
Successfully installed Appium-Python-Client-0.40
```

若最后的结果与以上示例相同,则表示安装成功。

9.1.7 Appium 连接模拟器和真机

在进行移动端自动化测试时,可以选择使用 SDK Manager 来安装指定版本的模拟器、第三方模拟器(夜神、逍遥等),或者使用真机来测试。本小节演示使用夜神模拟器和真机来连接测试。

1. 连接夜神模拟器

访问夜神官网,直接下载夜神模拟器即可,采用默认设置进行安装。下载完成后,需要将 Android SDK 中 platform-tools 目录下的 adb.exe 文件复制到夜神模拟器安装所在的 bin 目录下,并改名为 nox_adb.exe,将 bin 目录下原有的 nox_adb.exe 改名为 nox_adb.exe111。

启动夜神模拟器,查看设备是否连接成功。打开 cmd 命令提示符界面,输入"adb devices",示例如下:

```
C:\Users\23939>adb devices
List of devices attached
127.0.0.1:62001 device// 夜神的设备名称
```

若最后的结果与以上示例相同,则表示安装成功。

2. 连接真机

如果使用真机来测试,则需要配置以下信息。

(1)打开手机的 USB 调试模式,不同的手机打开调试模式的方式不同。一般都在开发者模式中打开。

(2)手机连接计算机。使用数据线连接手机与计算机,然后打开 cmd 命令提示符界面,输入"adb devices",查看连接是否成功。

可能会出现两个问题,一个是输入 adb 命令时提示不是内部命令或外部命令,那么需要下载一个 adb 工具包并存放在 C:\Windows\System32 目录下(注意,必须与 sdk 中的 adb 工具包版本一致);另一个是需要打开手机的 USB 调试模式并安装好驱动。

9.1.8 获取 APP 包名和 AppActivity

包名(PackageName)在 Android 系统中是 APP 的唯一标识,不同的 APP 可以有同样的名称,但它的包名不可以相同。

查看包名的方式有很多种,这里通过 Logcat 日志来查看包名。打开夜神模拟器,在模拟器中安装一个测试 apk 包,然后打开 cmd 命令提示符界面,输入"adb logcat -v time | findstr START",示例如下:

```
C:\Users\23939>adb logcat -v time | findstr START
03-26 20:09:34.041 W/PackageManager( 358): Unknown permission com.google.android.hangouts.START_HANGOUT in package com.google.android.gms
03-26 20:09:34.341 W/PackageManager( 358): Unknown permission com.google.android.hangouts.START_HANGOUT in package com.google.android.gms
03-26 20:09:41.241 I/ActivityManager( 358): START u0 {act=android.intent.action.MAIN cat=[android.intent.category.HOME] flg=0x10000000 cmp=com.vphone.launcher/.Launcher} from pid 0
03-26 20:09:49.898 D/AndroidRuntime( 957): >>>>>> AndroidRuntime START com.android.internal.os.RuntimeInit <<<<<<
03-26 20:13:23.808 I/ActivityManager( 358): START u0 {act=android.intent.action.MAIN cat=[android.intent.category.LAUNCHER] flg=0x10200000 cmp=io.appium.settings/.Settings bnds=[572,388][713,618] (has extras)} from pid 685
03-26 20:13:26.028 I/ActivityManager( 358): START u0 {act=com.android.systemui.recent.action.TOGGLE_RECENTS flg=0x10800000 cmp=com.android.systemui/.recent.RecentsActivity} from pid 533
03-26 20:13:26.908 I/ActivityManager( 358): START u0 {act=android.intent.action.MAIN cat=[android.intent.category.HOME] flg=0x10200000 cmp=com.vphone.launcher/.Launcher} from pid 533
03-26 20:15:33.498 I/ActivityManager( 358): START u0 {act=android.intent.action.
```

```
MAIN cat=[android.intent.category.LAUNCHER] flg=0x10200000 cmp=iflytek.testTech.
propertytool/.activity.BootActivity bnds=[290,388][431,618] (has extras)} from pid 685
03-26 20:15:35.048 I/ActivityManager( 358): START u0 {cmp=iflytek.testTech.
propertytool/.activity.HomeActivity} from pid 2256
```

单击模拟器界面上安装好的测试 apk 包。从上述返回的日志中可以看到 {cmp=iflytek.testTech.propertytool/.activity.HomeActivity}，cmp 等号后面的 iflytek.testTech.propertytool 值就是该测试 apk 的包名，.activity.HomeActivity 是启动 APP 包时需要的 Activity 信息。

知识拓展：Activity 是安卓应用程序的四大组件之一，负责管理安卓应用的用户界面。一个程序可以包含若干 Activity，每个 Acticity 负责展示一个用户界面。也就是说，在 Android 中每个界面都是一个 Activity，切换界面操作其实是多个不同的 Activity 之间的实例化操作。在 Android 中，Activity 的启动模式决定了 Activity 的启动运行方式。

9.1.9 Appium 第一个自动化脚本

启动 Appium 工具，单击 Appium 工具右上方的三角箭头，当变成正方形图标时，表明已经成功启动 Appium 服务。开启夜神模拟器，在模拟器上安装测试 apk 包，然后打开 cmd 命令提示符界面，输入 "adb install D:\android\apk\iTest.apk" [D:\android\apk 表示测试 apk（iTest.apk）本机所在目录位置]，示例如下：

```
C:\Users\23939>adb install D:\android\apk\iTest.apk
4664 KB/s (3281286 bytes in 0.687s)
        pkg: /data/local/tmp/iTest.apk
Success
```

此时，在夜神模拟器上会显示一个 iTest 图标，表明已经安装成功。打开 Pycharm 编辑器，在编辑器写入如下测试代码：

```python
import time  # 导入时间模块
from appium import webdriver  # 导入 WebDriver
desired_caps = {
                'platformName': 'Android',
                'platformVersion': '4.4.2',
                'deviceName': '127.0.0.1:62001',
                'appPackage': 'iflytek.testTech.propertytool',
                'appActivity': '.activity.BootActivity',
                'unicodeKeyboard': "True",# 支持输入中文
                'resetKeyboard': "True"
                }
driver = webdriver.Remote('http://127.0.0.1:4723/wd/hub',desired_caps)
```

当运行上述代码时，夜神模拟器界面上会出现 Appium 相关的标识图标，表示 Appium 自动化

测试环境彻底安装成功，如图 9.13 所示。

图 9.13　Appium 启动相关图标界面

下面针对 Desired_Capabilities 进行详解。Desired_Capabilities 是由多组 Key 和 Value 对象构成的参数集合。它的作用是负责启动服务端的参数设置，如设置测试平台、设备名、apk 包名，设置启动 Activity 等信息。

- platformName：表示使用哪种移动操作平台，是 iOS、Android 或 FirefoxOS 等。
- platformVersion：移动操作系统的版本号。
- deviceName：使用移动设备或模拟器的种类。http://127.0.0.1:4723/wd/hub 中的 127.0.0.1 表示本机地址；wd 可以理解为 WebDriver 的缩写；4723 表示 Appium 服务器端口号，可以在 Appium 设置中查看，如图 9.14 所示。

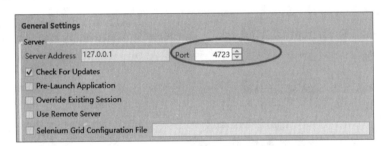

图 9.14　Appium 默认端口号界面

- appPackage：表示运行的 Android 应用的 Java 程序包，可以通过 adb 命令获取。
- appActivity：启动 APP 包时需要的 Activity 信息。

9.2　使用 Monitor 定位元素

不管是做 Web 端自动化测试还是 APP 端自动化测试，元素定位技术在自动化测试过程中都是非常重要的。在移动端领域常用来识别元素的工具有 uiautomator、Monitor 及 Appium Inspector。本节元素定位技巧演示是基于 Monitor 工具来识别元素的（适用于 Windows 和 Mac 平台）。

9.2.1　id 定位

（1）在夜神模拟器上安装一个测试 apk 包，单击登录到 APP 系统首页。进入 Android SDK 下的 tools 目录，找到 monitor.bat 批处理文件，双击运行，弹出如图 9.15 所示界面。

图 9.15　Monitor 工具首页界面

（2）单击图 9.15 所示界面上的图标，如果弹出如图 9.16 所示界面，则表明 Monitor 已经检测到夜神模拟器设备。

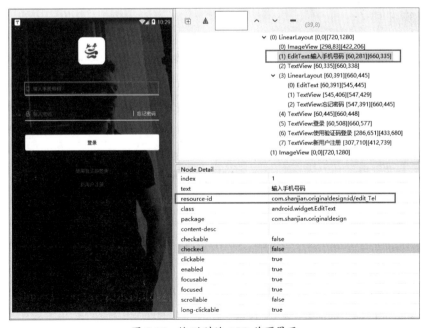

图 9.16　检测到的 APP 首页界面

在图 9.16 中，resource-id 对应的值 com.shanjian.originaldesign:id/edit_Tel 就是 id 的属性值。使用 ID 定位账号文本框，示例如下：

```python
from time import sleep
from appium import webdriver   # 导入 Appium 下的 WebDriver 模块
from selenium.webdriver.common.by import By   # 导入 By 类
desired_caps = {'platformName':'Android',
                'deviceName':'127.0.0.1:62001,
                'platformVersion':'3.8.3.1',
                'appPackage':'com.juyang.mall',
                'appActivity':'com.shanjian.juyang.activity.home.Activity_Home'
                }
desired_caps["unicodeKeyboard"] = "True"   # 使用 Unicode 编码方式发送字符串
desired_caps["resetKeyboard"] = "True"     # 将键盘隐藏起来
driver = webdriver.Remote('http://127.0.0.1:4723/wd/hub',desired_caps)   # 链接 APP
sleep(4)
driver.find_element(By.ID,'com.shanjian.originaldesign:id/edit_Tel').send_keys("18513600×××")
```

当输入用户名和密码时无法用 send_keys() 函数输入文本框中，是因为一输入就会打开软键盘，而软键盘上面的布局是九宫格，单击会一直出错。

desired_caps["unicodeKeyboard"] = "True" 和 desired_caps["resetKeyboard"] = "True" 的作用是以 unicodeKeyboard 的编码方式来发送字符串，并将键盘隐藏起来，然后就可以正常输入内容了。

9.2.2　name 定位

在使用 name 定位时，需要注意的是，Appium 1.5 以下旧版本可以通过 name 定位，新版本从 Appium 1.5+ 以后都不再支持 name 定位。本小节 Appium 工具安装的是 1.4 版本，故支持 name 定位。在移动端界面上，name 定位实际上就是 text 属性定位，示例如下：

```python
from time import sleep
from appium import webdriver   # 导入 Appium 下的 WebDriver 模块
from selenium.webdriver.common.by import By   # 导入 By 类
desired_caps = {'platformName':'Android',
                'deviceName':'127.0.0.1:62001,
                'platformVersion':'3.8.3.1',
                'appPackage':'com.juyang.mall',
                'appActivity':'com.shanjian.juyang.activity.home.Activity_Home'
                }
desired_caps["unicodeKeyboard"] = "True"   # 使用 Unicode 编码方式发送字符串
desired_caps["resetKeyboard"] = "True"     # 将键盘隐藏起来
driver = webdriver.Remote('http://127.0.0.1:4723/wd/hub',desired_caps)   # 链接 APP
```

```
sleep(4)
driver.find_element(By.Name,'输入手机号码').send_keys("18513600×××")
```

9.2.3　class 定位

class 在实际元素定位中非常常见，但笔者不建议经常使用 class 定位，因为 class 属性在一个页面重复的概率比较高。如果一定要使用 class 属性定位，则需要进行循环遍历和判断。如图 9.17 和图 9.18 所示，登录的账号文本框和密码文本框都有相同的 class 属性。

图 9.17　账号文本框的 class 属性值界面

图 9.18　密码文本框的 class 属性界面

由图 9.17 和图 9.18 可以发现，账号文本框和密码文本框的 class 属性值是相同的。针对这种情况，要使用 find_elements 复数的定位方式来定位元素，示例如下：

```
from time import sleep
from appium import webdriver  # 导入 Appium 下的 WebDriver 模块
from selenium.webdriver.common.by import By  # 导入 By 类
desired_caps = {'platformName':'Android',
                'deviceName':'127.0.0.1:62001,
```

```
                'platformVersion':'3.8.3.1',
                'appPackage':'com.juyang.mall',
                'appActivity':'com.shanjian.juyang.activity.home.Activity_Home'
               }
desired_caps["unicodeKeyboard"] = "True"   # 使用 Unicode 编码方式发送字符串
desired_caps["resetKeyboard"] = "True"     # 将键盘隐藏起来
# class_name 定位
buttons = driver.find_elements_by_class_name("android.widget.EditText")
# 通过 for 循环遍历
for button in range(len(buttons)):
    data = ['18513600×××','123456']
    element = driver.find_elements_by_class_name('android.widget.EditText')[button]
    element.send_keys(data[button])
```

先通过 find_elements() 方法判断当前页面中 class_name 值的个数,然后 for 循环依次将遍历的数值通过索引作用到每个操作中(如输入账号和输入密码)。

9.2.4 xpath 定位

xpath 定位语法非常丰富,包括属性定位、层级定位及文本定位等方法。本小节使用 xpath 来定位登录账号和密码文本框,示例如下:

```
...
desired_caps = {'platformName':'Android',
                'deviceName':'127.0.0.1:62001,
                'platformVersion':'3.8.3.1',
                'appPackage':'com.juyang.mall',
                'appActivity':'com.shanjian.juyang.activity.home.Activity_Home'
               }
desired_caps["unicodeKeyboard"] = "True"   # 使用 Unicode 编码方式发送字符串
desired_caps["resetKeyboard"] = "True"     # 将键盘隐藏起来
# 输入账号
driver.find_element_by_xpath("//android.widget.EditText[@resource-id='com.shanjian.originaldesign:id/edit_Tel']").send_keys('18513600×××')
# 输入密码
driver.find_element_by_xpath("//android.widget.EditText[@resource-id='com.shanjian.originaldesign:id/edit_Pwd']").send_keys('18513600×××')
# 单击 "登录" 按钮
driver.find_element_by_xpath("//android.widget.TextView[@text=' 登录 ']").click()
```

xpath 定位中的 android.widget.EditText 表示文本框(账号和密码)的 class 属性,[] 中可以存放 id、name(text=' 登录 ')等属性值。

9.2.5 accessibilty_id 定位

accessibility_id 定位其实是通过 Monitor 工具来查看 content-desc 属性，如图 9.19 所示。

Node Detail	
index	5
text	登录
resource-id	com.shanjian.originaldesign:id/tv_LoginBtn
class	android.widget.TextView
package	com.shanjian.originaldesign
content-desc	
checkable	false
checked	false
clickable	true
enabled	true
focusable	false
focused	false
scrollable	false
long-clickable	false

图 9.19 content-desc 属性界面

图 9.19 中的定位，如果看到 content-desc 属性有值，就可以通过 accessibility_id 来定位，格式为 driver.find_element_by_accessibility_id（content-desc 的值）。

9.2.6 android_uiautomator 定位

android_uiautomator 是通过 Android 自带的 Android UIAutomator 类库去查找元素。android_uiautomator 的元素定位其实与 Appium 的定位一样，或者说它比 Appium 的定位方式更加多也更加适用，示例如下：

```
desired_caps = {
            'platformName': 'Android',
            'platformVersion': '5.0.0.0',
            'deviceName': '127.0.0.1:62001',
            'appPackage': 'com.shanjian.originaldesign',
            'appActivity': '.activity.other.Activity_In',
            }
driver = webdriver.Remote('http://127.0.0.1:4723/wd/hub',desired_caps)
sleep(10)
# 用于清除历史记录
driver.find_element_by_id("com.shanjian.originaldesign:id/edit_Tel").clear()
driver.find_element_by_android_uiautomator('newUiSelector().text("输入手机号码")').send_keys("18513600×××")
driver.find_element_by_id("com.shanjian.originaldesign:id/edit_Pwd").send_keys("123456")
driver.find_element_by_android_uiautomator('new UiSelector().text("登录")').click()
```

注意：使用 android_uiautomator 定位时，new UiSelect() 方法中间要有空格，并且最外层使用单引号，里面使用双引号，否则会报如下错误：

```
selenium.common.exceptions.WebDriverException: Message: The requested resource
could not be found,or a request was received using an HTTP method that is not
supported by the mapped resource.
```

android_uiautomator 还支持 text 模糊定位。通过 text 部分信息就能够进行定位，示例如下：

```
ele = self.driver.find_element_by_android_uiautomator('new UiSelector().textContains(
    "请输入手")')
ele.send_keys("123456")
```

在上面的代码中只是将 .text() 方法变成了 .textContains() 方法，在方法中传入模糊的数据即可。此外，android_uiautomator 还支持 textMatches 正则匹配查找，该方法也是通过 text 的属性来进行正则匹配，示例如下：

```
ele = self.driver.find_element_by_android_uiautomator('new UiSelector().textMatches(
    "^请输入手.*")')
ele.send_keys("123")
```

不仅如此，android_uiautomator 还支持 resourceId 属性定位，resourceId 定位和 Appium 封装好的 id 定位是一样的，只是这里将写法变成了 android_uiautomator 的写法而已，示例如下：

```
ele = driver.find_element_by_android_uiautomator('new UiSelector().resourceId(
    "com.shanjian.originaldesign:id/edit_Tel")')
ele.send_keys('234')
```

9.3 Native App 实战

本节结合 Android 操作平台演示 Appium 在 Native App 中的应用。Native App 是一种基于智能手机本地操作系统，如 iOS、Android 和 WP，并使用原生程式编写运行的第三方应用程序，也称为本地 APP。其一般使用的开发语言为 Java、C++ 和 Objective-C 等。

9.3.1 模拟键盘事件

在元素定位时，有些情况下可能会遇到定位的元素不可单击，需要单击软键盘中的"发送"按钮才能提交。这时就需要模拟键盘输入了，通过发送键盘事件来达到定位效果。Appium 中提供了 pressKeyCode(AndroidKeyCode) 方法来模拟键盘输入。Android KeyCode 参考表如下（只限 Android 操作系统）：

电话键
KEYCODE_CALL 拨号键 5
KEYCODE_ENDCALL 挂机键 6
KEYCODE_HOME 按键 Home 3
KEYCODE_MENU 菜单键 82
KEYCODE_BACK 返回键 4
KEYCODE_SEARCH 搜索键 84
KEYCODE_CAMERA 拍照键 27
KEYCODE_FOCUS 拍照对焦键 80
KEYCODE_POWER 电源键 26
KEYCODE_NOTIFICATION 通知键 83
KEYCODE_MUTE 话筒静音键 91
KEYCODE_VOLUME_MUTE 扬声器静音键 164
KEYCODE_VOLUME_UP 音量增加键 24
KEYCODE_VOLUME_DOWN 音量减小键 25
控制键
KEYCODE_ENTER 回车键 66
KEYCODE_ESCAPE ESC键 111
KEYCODE_DPAD_CENTER 导航键 确定键 23
KEYCODE_DPAD_UP 导航键 向上 19
KEYCODE_DPAD_DOWN 导航键 向下 20
KEYCODE_DPAD_LEFT 导航键 向左 21
KEYCODE_DPAD_RIGHT 导航键 向右 22
KEYCODE_MOVE_HOME 光标移动到开始键 122
KEYCODE_MOVE_END 光标移动到末尾键 123
KEYCODE_PAGE_UP 向上翻页键 92
KEYCODE_PAGE_DOWN 向下翻页键 93
KEYCODE_DEL 退格键 67
KEYCODE_FORWARD_DEL 删除键 112
KEYCODE_INSERT 插入键 124
KEYCODE_TAB Tab键 61
KEYCODE_NUM_LOCK 小键盘锁 143
KEYCODE_CAPS_LOCK 大写锁定键 115
KEYCODE_BREAK Break/Pause键 121
KEYCODE_SCROLL_LOCK 滚动锁定键 116
KEYCODE_ZOOM_IN 放大键 168
KEYCODE_ZOOM_OUT 缩小键 169
组合键
KEYCODE_ALT_LEFT Alt+Left
KEYCODE_ALT_RIGHT Alt+Right
KEYCODE_CTRL_LEFT Control+Left
KEYCODE_CTRL_RIGHT Control+Right
KEYCODE_SHIFT_LEFT Shift+Left
KEYCODE_SHIFT_RIGHT Shift+Right
基本键
KEYCODE_0 按键 '0' 7

KEYCODE_1	按键	'1'	8
KEYCODE_2	按键	'2'	9
KEYCODE_3	按键	'3'	10
KEYCODE_4	按键	'4'	11
KEYCODE_5	按键	'5'	12
KEYCODE_6	按键	'6'	13
KEYCODE_7	按键	'7'	14
KEYCODE_8	按键	'8'	15
KEYCODE_9	按键	'9'	16
KEYCODE_A	按键	'A'	29
KEYCODE_B	按键	'B'	30
KEYCODE_C	按键	'C'	31
KEYCODE_D	按键	'D'	32
KEYCODE_E	按键	'E'	33
KEYCODE_F	按键	'F'	34
KEYCODE_G	按键	'G'	35
KEYCODE_H	按键	'H'	36
KEYCODE_I	按键	'I'	37
KEYCODE_J	按键	'J'	38
KEYCODE_K	按键	'K'	39
KEYCODE_L	按键	'L'	40
KEYCODE_M	按键	'M'	41
KEYCODE_N	按键	'N'	42
KEYCODE_O	按键	'O'	43
KEYCODE_P	按键	'P'	44
KEYCODE_Q	按键	'Q'	45
KEYCODE_R	按键	'R'	46
KEYCODE_S	按键	'S'	47
KEYCODE_T	按键	'T'	48
KEYCODE_U	按键	'U'	49
KEYCODE_V	按键	'V'	50
KEYCODE_W	按键	'W'	51
KEYCODE_X	按键	'X'	52
KEYCODE_Y	按键	'Y'	53
KEYCODE_Z	按键	'Z'	54

下面使用press_keycode()方法来模拟键盘操作数字、模拟回车按键等，示例如下：

```
from appium import webdriver
from time import sleep
desired_caps = {
            'platformName': 'Android',
            'platformVersion': '5.0.0.0',
            'deviceName': '127.0.0.1:62001',
            'appPackage': 'com.shanjian.originaldesign',
            'appActivity': '.activity.other.Activity_In'
```

```python
        }
driver = webdriver.Remote('http://127.0.0.1:4723/wd/hub',desired_caps)
    def testAdd(self):
        sleep(3)
        # 按下键盘 1
        self.dr.press_keycode(8)
        # 按下键盘 0
        self.dr.press_keycode(7)
        # 定位一个元素
        addoperation = self.dr.find_element(By.ID,"plus")
        # 操作一个元素
        addoperation.click()
        sleep(1)
        # 按下键盘 5
        self.dr.press_keycode(12)
        # 定位一个元素
        equal = self.dr.find_element_by_accessibility_id("equals")
        # 操作一个元素
        equal.click()
        # 模拟回车按键 66
        self.dr.press_keycode(66)
        sleep(3)
        # Exception 异常处理
        try:
            result = self.dr.find_element(By.CLASS_NAME,"android.widget.EditText")
            value = result.text
            # Verify 断言，验证操作的结果
            self.assertEqual("103",value)
        except Exception as msg:
            print(" 程序出现了异常：",msg)
    def tearDown(self):
        self.dr.quit()
if __name__ == '__main__':
    unittest.main()
```

9.3.2 滑动封装实战

滑动应用在移动端自动化测试过程中很常见，可以分为向上滑动、向下滑动、向左滑动和向右滑动等。本小节通过实战案例来演示滑动实战。官方给出的案例如下：

```
swipe(self,start_x,start_y,end_x,end_y,duration=None)
...
解释：
int start_x——开始滑动的 x 坐标；
```

```
int start y ——开始滑动的 y 坐标；
int end x ——结束点 x 坐标；
int end y ——结束点 y 坐标；
duration ——滑动时间（默认 5ms）
'''
```

屏幕左上角为起点，坐标为 (0，0)，起点往右为 X 轴，起点以下为 Y 轴。下面介绍 swipe 滑动在 Appium 自动化测试中的实战应用，以某商城测试 APP 为例，示例如下：

```python
from time import sleep
from appium import webdriver
import unittest
class MyTestCase(unittest.TestCase):
    def setUp(self):
        desired_caps = {}
        desired_caps['platformName'] = 'Android'
        desired_caps['platformVersion'] = '5.0.0.0'       # Android 操作系统的版本
        desired_caps['deviceName'] = '127.0.0.1:62001'    # 设备名称
        desired_caps['appPackage'] = 'com.juyang.mall'    # APP 的包名
        desired_caps['appActivity'] = 'com.shanjian.juyang.activity.home.Activity_Home'
        # 指定脚本要把内容发送到的地址及端口
        self.driver = webdriver.Remote('http://127.0.0.1:4723/wd/hub',desired_caps)
    def test_swipe(self):
        sleep(5)
        # 向上滑动
        self.swipeUp(self.driver)
        # 向下滑动
        sleep(5)
        self.swipeDown(self.driver)
    def swipeUp(self,driver,t=500):
        '''向上滑动'''
        l = self.driver.get_window_size()
        print(l)
        print(type(l))             # 字典类型
        x1 = l['width'] * 0.5      # x 坐标
        y1 = l['height'] * 0.75    # 起始 y 坐标
        y2 = l['height'] * 0.25    # 终点 y 坐标
        self.driver.swipe(x1,y1,x1,y2,t)
    def swipeDown(self,self,driver,t=500):
        '''向下滑动'''
        l = self.driver.get_window_size()
        x1 = l['width'] * 0.5      # x 坐标
        y1 = l['height'] * 0.25    # 起始 y 坐标
        y2 = l['height'] * 0.75    # 终点 y 坐标
        self.driver.swipe(x1,y1,x1,y2,t)
    def swipLeft(self,driver,t=500):
```

```python
        '''向左滑动屏幕'''
        l = self.driver.get_window_size()
        x1 = l['width'] * 0.75
        y1 = l['height'] * 0.5
        x2 = l['width'] * 0.25
        self.driver.swipe(x1,y1,x2,y1,t)
    def swipRight(self,driver,t=500):
        '''向右滑动屏幕'''
        l = self.driver.get_window_size()
        x1 = l['width'] * 0.25
        y1 = l['height'] * 0.5
        x2 = l['width'] * 0.75
        self.driver.swipe(x1,y1,x2,y1,t)
    def tearDown(self):
        pass
if __name__ == "__main__":
    unittest.main(verbosity=2)
```

通过 get_window_size() 方法获取手机屏幕尺寸长（height）和宽（width），水平滑动时，X 轴变化，Y 轴不变化；垂直滑动时正好相反，Y 轴变化，X 轴不变化。t 是滑动的持续时间，默认可以不写。

9.3.3 多点触控实战

多点触控实际上是触摸动作的集合，本小节演示多点触控中的 tap() 方法应用实例。tap 表示触摸，与 click 单击的效果是一样的，只不过其是通过坐标来找到被定位的元素的，示例如下：

```python
import unittest
from time import sleep
from appium import webdriver
class MyTestCase(unittest.TestCase):
    # 脚本初始化，获取操作实例
    def setUp(self):
        desired_caps = {}
        desired_caps['platformName'] = 'Android'
        desired_caps['platformVersion'] = '5.0.0.0'        # Android 操作系统的版本
        desired_caps['deviceName'] = '127.0.0.1:62001'     # 设备名称
        desired_caps['appPackage'] = 'com.juyang.mall'     # APP 的包名
        # APP 的 Activity 名称
        desired_caps['appActivity'] = 'com.shanjian.juyang.activity.home.Activity_Home'
# 使用 Appium 的 Unicode 键盘
desired_caps['unicodeKeyboard'] = 'True'
        desired_caps['restKepassyboard'] = 'True'  # True -- 测试执行完成后，恢复原来的输入法
        self.dr = webdriver.Remote('http://127.0.0.1:4723/wd/hub',desired_caps)
```

```
    def testMoreAPIs(self):
        sleep(4)
        # tap([(元组1),(元组2),(元组3)])
        self.dr.tap([(143,1199),(288,1280)])
        sleep(4)
    def tearDown(self):
        self.dr.close_app()
if __name__ == '__main__':
    unittest.main()
```

通过上述代码实例，不难发现在使用 tap() 方法时，是以元组的形式来存储元素的坐标，如([(143,1199),(288,1280)])。

9.3.4 自动化异常截图

不管是在手工测试还是自动化测试过程中，如果发现问题（Bug），保留证据（截图、抓包、查看日志等）是测试人员最强有力的凭证。Appium 中提供了多种截图方法来自动生成截图，比较常用的是 get_screenshot_as_file() 方法，示例如下：

```
from appium import webdriver
from time import sleep
from selenium.webdriver.common.by import By   # 导入By类
import time,os
desired_caps = {
                'platformName': 'Android',
                'platformVersion': '5.0.0.0',
                'deviceName': '127.0.0.1:62001',
                'appPackage': 'com.shanjian.originaldesign',
                'appActivity': '.activity.other.Activity_In'
                }
driver = webdriver.Remote('http://127.0.0.1:4723/wd/hub',desired_caps)
sleep(10)
driver.find_element_by_id('com.shanjian.originaldesign:id/edit_Tel').send_keys('18513600×××')
driver.find_element_by_id('com.shanjian.originaldesign:id/edit_Pwd').send_keys('123456')
driver.find_element(By.ID,'com.shanjian.originaldesign:id/edit_Tel').click()
def insert_img(driver,filename):
    ''' 截图函数'''
    img_folder = os.path.dirname(__file__)
    now = time.strftime("%Y-%m-%d_%H_%M_%S")
    files = img_folder + '\\' + now + filename
    driver.get_screenshot_as_file(files)
if __name__ == '__main__':
    insert_img(driver,'error.png')
```

调用 imsert_img() 方法时，传入驱动 driver 和截图名称即可。

注意：截图名称的扩展名必须是以 .png 格式结尾的，否则会产生如下报错信息：

```
C:\Python36\lib\site-packages\selenium\webdriver\remote\webdriver.py:1031:
UserWarning: name used for saved screenshot does not match file type. It should end
with a `.png` extension
  "type. It should end with a `.png` extension",UserWarning)
```

time.strtime() 函数返回的是一个时间字符串，表示当地时间，目的是用当地时间来区分不同的截图名称。截图效果如图 9.20 所示。

图 9.20　自动化异常截图

9.4　Appium 完整脚本实战

本节结合实战案例从自动化需求分析入手，系统地阐述自动化测试用例设计和脚本封装等技巧。在演示过程中会涉及前面章节所学习的元素定位、常用 WebDriver 操作及 unittest 单元测试框架等知识。通过本节的学习，读者可以在拿到测试需求之后，顺利地开展移动端自动化测试任务。

9.4.1　测试需求分析

本小节以简书 APP 项目为例进行演示，如图 9.21 所示。

图 9.21　简书登录界面

因为登录功能一般是比较稳定的，界面元素很少会经常变动，且系统一些主要功能都是依赖于登录设计开发的，所以从登录功能入手是一个不错的开始。同样地，笔者也建议测试人员在最开始从事移动端自动化测试时，优先选择登录功能入手和分析。

从登录界面来看，最常见的是登录账号的文本框、登录密码文本框、登录按钮等控件。下面使用 Android SDK 目录下的 Monitor 工具来定位元素，如图 9.22 所示。

图 9.22　Monitor 工具识别简书登录界面

9.4.2　测试用例设计

登录是经常要使用到的功能，虽然有时不登录也可以操作系统，但操作到后面的一些环境时都会验证是否登录。最简单的案例就是淘宝购物，如果不登录淘宝系统，就无法购买商品和下单。鉴于以上情况，可以把登录封装为一个独立的功能。

此外，在登录时，可能会存在很多种正常和异常的登录情况。例如，输入正确的用户名和密码，登录系统成功；输入正确的用户名和错误的密码，登录失败等。当登录成功后，需要获取登录成功后系统返回的标识，即断言。简单来说，就是正常的测试数据对应正常的测试结果，异常的测试数据对应异常的测试结果。下面对简书 APP 登录界面进行自动化测试用例设计，如图 9.23 所示。

用例编号	模块	前提条件	操作步骤	预期结果	实际结果
Login_001	登录	系统登录且正常运行	1.检查用户名输入框id为com.jianshu.haruki:id/et_account 2.检查账号文本框id为com.jianshu.haruki:id/et_verification_code_or_password 3.检查登录按钮元素id为com.jianshu.haruki:id/tv_login 4.在用户名输入框id为com.jianshu.haruki:id/et_account输入账号：xxxx 5.在密码输入框com.jianshu.haruki:id/et_verification_code_or_password输入密码：xxxx 6.单击id为com.jianshu.haruki:id/tv_login登录按钮 7.验证是否登录成功	账号输入框状态正常 密码输入框状态正常 登陆按钮状态正常 账号输入正确 密码输入正确 单击登录按钮，页面跳转首页 登录成功	

图 9.23　自动化测试用例设计界面

把登录需要用到的每一个操作对应的元素写在了 Excel 表中，方便引用。

9.4.3 测试脚本编写

根据图 9.23，下面编写登录功能的自动化测试用例设计。根据登录功能，设计 3 组测试数据，每组测试数据对应不同的测试结果。测试用例分别如下。

（1）输入正确的用户名和正确的密码，登录成功。

（2）输入用户名为空和正确的密码，提示：输入的用户名/密码不正确。

（3）输入正确的用户名和密码为空，提示：输入的用户名/密码不正确。

登录脚本代码示例如下：

```python
from appium import webdriver
import unittest,time
from selenium.webdriver.common.by import By   # By 类定位
class TestItest(unittest.TestCase):
    def setUp(self):
        desired_caps = {
                        'platformName': 'Android',
                        'platformVersion': '4.4.2',
                        'deviceName': '127.0.0.1:62001',
                        # 简书 APP 包名
                        'appPackage': '= com.jianshu.haruki',
                        'appActivity': 'com.baiji.jianshu.ui.splash.SplashScreenActivity',
                        # 支持输入中文
                        'unicodeKeyboard': "True",
                        'resetKeyboard': "True"
                        }
        self.driver = webdriver.Remote('http://127.0.0.1:4723/wd/hub',desired_caps)
        # 等待 APP 首页出现
        self.driver.wait_activity('.activity.HomeActivity',20)
    def loginsys(self,username,password):
        '''登录流程'''
        self.driver.find_element(By.ID,"com.jianshu.haruki:id/et_account").send_keys('username')
        self.driver.find_element(By.ID,"com.jianshu.haruki:id/et_verification_code_or_password").send_keys('password')
        self.driver.find_element(By.ID,"com.jianshu.haruki:id/tv_login").click()
        time.sleep(3)
    def exceptRes(self):
        '''对测试结果进行断言'''
        try:
            text = self.driver.find_element(By.ID,'com.jianshu.haruki:id/tv_switch_login_mode').text
            return text
    def test_user_passwd_success(self):
        '''输入正确的用户名和密码，登录成功'''
```

```python
        self.loginsys('18513600×××','213456')
        self.assertEqual('登录成功',self.exceptRes())
    def test_user_is_null(self):
        ''' 输入用户名为空,提示输入用户名/密码不正确 '''
        self.loginsys('','11123')
        self.assertEqual('输入用户名/密码不正确',self.exceptRes())

    def test_password_is_null(self):
        ''' 输入密码为空,提示输入用户名/密码不正确 '''
        self.loginsys('13217690×××','')
        self.assertEqual('输入用户名/密码不正确',self.exceptRes())
    def tearDown(self):
        self.driver.quit()
if __name__ == '__main__':
    unittest.main(verbosity=2)
```

将连接 Appium 服务配置信息 desired_caps 存放在 setUp() 方法中,方便每个测试用例在执行前都会先执行 setUp() 方法中的连接操作。之后,将登录流程封装到 loginsys() 方法中。在用例层中分别设计了 3 条测试用例来验证登录功能。exceptRes() 方法用于对登录后的文本结果断言。

9.4.4 测试结果及分析

在实际项目中,自动化测试用例跑完以后,需要对测试结果进行收集和整理,可以通过收集日志信息或测试报告来查看测试结果。

Python 中提供了 HTMLTestRunner.py 来生成测试报告,下载 HTMLTestRunner.py 后,直接放到 Python 安装目录的 Lib 文件夹中。继续对 9.4.3 小节中的代码案例进行改进,增加如下代码:

```python
from appium import webdriver
import unittest,time
from selenium.webdriver.common.by import By   # By 类定位
from HTMLTestRunner import HTMLTestRunner
class JianBook(unittest.TestCase):
    def setUp(self):
        desired_caps = {
                        'platformName': 'Android',
                        'platformVersion': '4.4.2',
                        'deviceName': '127.0.0.1:62001',
                        # 简书 APP 包名
                        'appPackage': '= com.jianshu.haruki',
                        'appActivity': 'com.baiji.jianshu.ui.splash.SplashScreenActivity',
                        # 支持输入中文
                        'unicodeKeyboard': "True",
                        'resetKeyboard': "True"
```

```python
            }
        self.driver = webdriver.Remote('http://127.0.0.1:4723/wd/hub',desired_caps)
        # 等待 APP 首页出现
        self.driver.wait_activity('.activity.HomeActivity',20)
    def loginsys(self,username,password):
        '''登录流程'''
        self.driver.find_element(By.ID,"com.jianshu.haruki:id/et_account").send_keys('username')
        self.driver.find_element(By.ID,"com.jianshu.haruki:id/et_verification_code_or_password").send_keys('password')
        self.driver.find_element(By.ID,"com.jianshu.haruki:id/tv_login").click()
        time.sleep(3)
    def exceptRes(self):
        '''对测试结果进行断言'''
        try:
            text = self.driver.find_element(By.ID,'com.jianshu.haruki:id/tv_switch_login_mode').text
            return text
    def test_user_passwd_success(self):
        '''输入正确的用户名和密码，登录成功'''
        self.loginsys('18513600×××','213456')
        self.assertEqual('登录成功',self.exceptRes())
    def test_user_is_null(self):
        '''输入用户名为空，提示输入用户名/密码不正确'''
        self.loginsys('','11123')
        self.assertEqual('输入用户名/密码不正确',self.exceptRes())
    def test_password_is_null(self):
        '''输入密码为空，提示输入用户名/密码不正确'''
        self.loginsys('13217690×××','')
        self.assertEqual('输入用户名/密码不正确',self.exceptRes())
    def tearDown(self):
        self.driver.quit()
if __name__ == '__main__':
    # 定义测试报告所在目录
    report_dir = r'...\result.html'
    # 定义文件报告权限，wb 表示读写权限
    re_open = open(report_dir,'wb')
    # 根据给定的测试类，获取其中的所有测试方法，并返回一个测试套件
    suite = unittest.TestLoader().loadTestsFromTestCase(JianBook)
    runner = HTMLTestRunner(
                    stream=re_open,
                    title=u'Appium 自动化测试实战_jianshubookPro',
                    description=u'测试报告')
    # 执行测试套件
    runner.run(suite)
```

如上述代码所示，将 3 条测试用例的运行结果写入当前目录的 result.html 文件中，测试结果如图 9.24 所示。

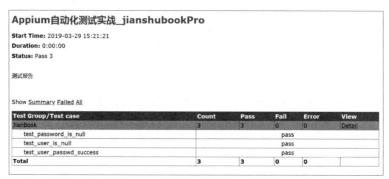

图 9.24　Appium 测试报告

 ## Appium 常见问题

在使用 Appium 进行自动化测试时，会遇到各种各样的问题。笔者为读者列举了一些比较常见的问题解决思路，方便读者参考借鉴。关于更多实际开发中遇到的问题，读者可以参考其他资料。

有些情况下启动 uiautomatorviewer.bat 会报错，如图 9.25 所示。

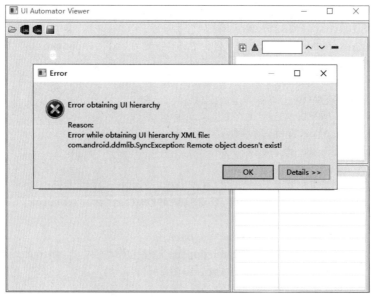

图 9.25　UI Automator Viewer 启动报错界面

上述报错的解决办法如下。

（1）打开 cmd 命令提示符界面，输入"adb root"，然后重新打开，如果还是不行，可多试几次。

（2）检查 JDK 版本是否为 1.8 或 1.8 以下，因为版本为 1.9 的 JDK 不兼容 UI Automator Viewer。

（3）打开 cmd 命令提示符界面，先输入"adb shell"，然后输入"su root"，接着对 tmp 文件赋权，输入"chmod 777 /data/local/tmp"，再次启动 UI Automator Viewer。

A new session could not be create 报错的解决办法：在 Desired_caps 字典中添加 AppWaitActivity 关键字，该关键字为 APP 首页的 Activity，可通过 Appium 获取，AppWaitActivity=com.shanjian.juyang.activity.home.Activity_Home，如图 9.26 所示。

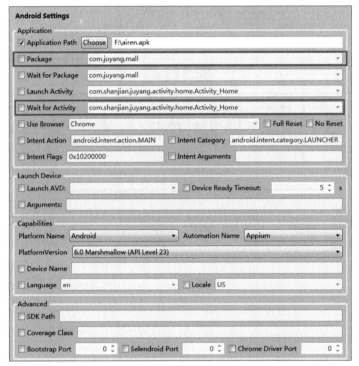

图 9.26　Android Settings 界面

随着自动化测试案例越来越丰富，到后期就不得不考虑自动化测试的框架开发了，这里面会涉及更多的开发知识，如配置文件读取、日志文件管理、更多代码设计模式等。也希望读者可以在本节知识的基础之上，继续深入学习 Appium 测试框架并将其应用到项目中，在完善自身技能的同时为企业创造效益。

第10章
Git版本控制工具实战

Git 是一个用于 Linux 内核开发的开源的分布式版本控制系统，可以有效、高速地处理从很小到非常大的项目版本管理。与常用的集中式版本控制工具 Subversion 等不同，Git 采用了分布式版本库的方式。

分布式相比于集中式的最大区别在于开发者可以提交到本地，每个开发者通过克隆（git clone），在本地机器上复制一个完整的 Git 仓库，并且工作时不需要联网。本章进行 Git 相关知识的学习。

10.1 搭建 Git 环境

Git 目前可以在 Linux、UNIX、Mac 和 Windows 平台上安装并运行。本节主要演示在 64 位 Windows 平台上安装 Git。Git 的官方下载地址为 https://git-scm.com/download，注意，根据系统版本选择对应的版本下载并安装。下载完成后，双击文件进行安装，选择安装目录，如图 10.1 所示。

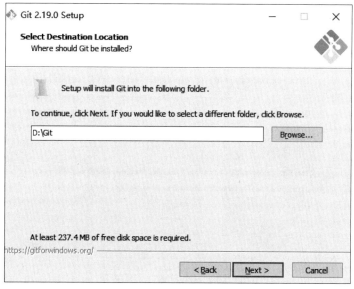

图 10.1　Git 安装目录界面

单击"Next"按钮，选中安装 Git 所需的组件（Git Bash Here 表示支持命令行模式，Git GUI Here 表示支持图形化界面），如图 10.2 所示。

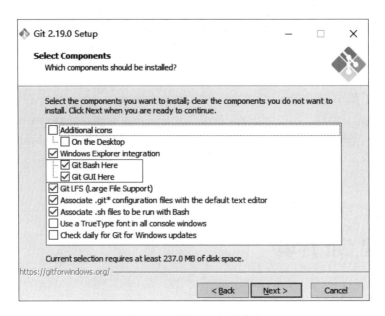

图 10.2　设置 Git 组件界面

单击"Next"按钮,选中"Use Git from Git Bash only"单选按钮,如图 10.3 所示。

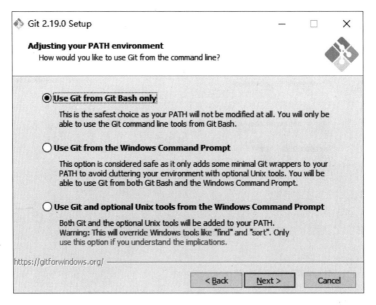

图 10.3　命令行环境设置界面

单击"Next"按钮,接下来的四步采用默认选项,直接进行下一步操作。最后单击"Finish"按钮,如图 10.4 所示。

图 10.4　安装完成界面

检查 Git 环境是否安装成功，可回到桌面右击，在弹出的快捷菜单中查看是否包含 Git 相关的两个模式，如果有，则表示安装成功，如图 10.5 所示。

图 10.5　Git 相关的两个模式

Git 基本操作

本节学习 Git 常见的基本操作，如创建版本库、添加文件、文件跟踪管理及版本回退等。

10.2.1 创建版本库

Git 安装完成后,要在本地计算机上创建一个版本库,可以把版本库简单理解为存放众多文件的一个目录管理中心。在这个目录中,所有文件都可以被 Git 管理,包括每个文件的修改、删除等。

Git 是基于 Linux 内核开发的版本控制工具,所以支持 Linux 命令操作。在本地 D 盘创建一个 mygit 目录,示例如下:

```
23939@DESKTOP-80LK128 MINGW64 /d
$ mkdir mygit
23939@DESKTOP-80LK128 MINGW64 /d
$ cd  mygit/
23939@DESKTOP-80LK128 MINGW64 /d/mygit
$ pwd
/d/mygit
23939@DESKTOP-80LK128 MINGW64 /d/mygit
$
```

mkdir 命令用于创建一个目录,pwd 表示显示当前目录所在的绝对路径。在创建目录时,确保目录名称中不要包含中文。此时在 D 盘下会生成一个 mygit 的目录。

创建 mygiy 目录后,使用 git init 命令把目录变成 Git 可以管理的仓库,示例如下:

```
23939@DESKTOP-80LK128 MINGW64 /d/mygit
$ git init
Initialized empty Git repository in D:/mygit/.git/
23939@DESKTOP-80LK128 MINGW64 /d/mygit (master)
$
```

在 D 盘 mygit 目录下多了一个 .git 的文件夹,这个目录是 Git 用来跟踪管理版本库的,不要删除或修改目录中的文件,如图 10.6 所示。

图 10.6 初始化 Git 版本库界面

10.2.2 添加文件

在 mygit 目录下创建一个 .txt 格式的文件,可以手动创建 .txt 文件,也可以在 Git 中使用 touch

命令来创建文件，示例如下：

```
23939@DESKTOP-80LK128 MINGW64 /d/mygit (master)
$ touch studygit.txt
23939@DESKTOP-80LK128 MINGW64 /d/mygit (master)
$ ls
studygit.txt
23939@DESKTOP-80LK128 MINGW64 /d/mygit (master)
$
```

使用 vi 命令进入 studygit.txt 文件中新增一些内容，代码如下：

```
23939@DESKTOP-80LK128 MINGW64 /d/mygit (master)
$ vi studygit.txt
```

在 studygit.txt 文件中新增如下内容：

```
hello,git!
This is my first learn git!!!
```

需要说明的是，studygit.txt 文本必须放在 mygit 目录下，因为 mygit 是 Git 的仓库，如果放在其他目录下，Git 是无法找到的。

使用 git add 命令将 studygit.txt 文件添加到仓库，代码如下：

```
23939@DESKTOP-80LK128 MINGW64 /d/mygit (master)
$ git add studygit.txt
23939@DESKTOP-80LK128 MINGW64 /d/mygit (master)
$
```

出现上述内容，则说明添加文件到仓库是成功的。有时执行 git add studygit.txt 命令时会出现报错信息，如图 10.7 所示。

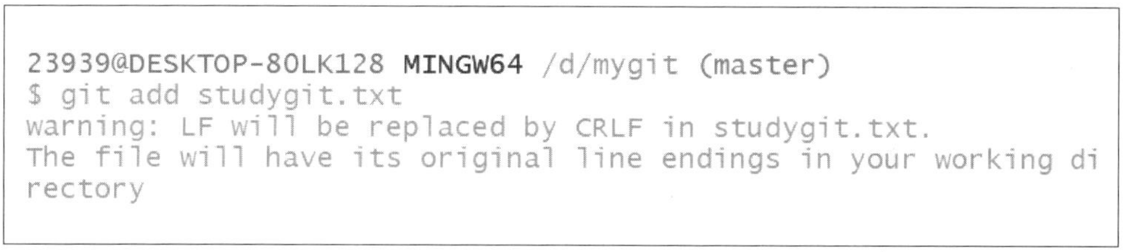

```
23939@DESKTOP-80LK128 MINGW64 /d/mygit (master)
$ git add studygit.txt
warning: LF will be replaced by CRLF in studygit.txt.
The file will have its original line endings in your working di
rectory
```

图 10.7　执行 git add 命令报错界面

报错原因是 Windows 中的换行符为 CRLF，而 Linux 下的换行符为 LF（使用 Git 命令行 Git Bash，实际上就是相当于 Linux 环境），所以在执行 git add studygit.txt 操作时会出现这个错误提示。解决办法如下：

```
23939@DESKTOP-80LK128 MINGW64 /d/mygit
$ rm -rf .git
23939@DESKTOP-80LK128 MINGW64 /d/mygit
$ git config --global core.autocrlf false
23939@DESKTOP-80LK128 MINGW64 /d/mygit
$ git init
Initialized empty Git repository in D:/mygit/.git/
23939@DESKTOP-80LK128 MINGW64 /d/mygit (master)
$ git add studygit.txt
23939@DESKTOP-80LK128 MINGW64 /d/mygit (master)
$
```

使用 rm -rf 命令删除 .git 目录，git config --global core.autocrlf false 表示禁用自动转换，最后使用 git init 命令重新初始化 mygit 目录。

使用 git commit -m 命令将文件提交到仓库，示例如下：

```
23939@DESKTOP-80LK128 MINGW64 /d/mygit (master)
$ git commit -m '第一次提交 studygit.txt'
[master (root-commit) 0fa8d45] 第一次提交 studygit.txt
 1 file changed,2 insertions(+)
 create mode 100644 studygit.txt
```

在 git commit -m 后输入的是本次提交的说明，可以输入自定义的内容。git commit 命令执行成功后返回的信息中，1 file changed 表示有一个文件被改动（新添加的 studygit.txt 文件），2 insertions 表示插入了两行内容（studygit.txt 有两行内容）。

注意：使用 Git 添加文件要执行两步操作，即 add 操作和 coomit 操作。因为每次 add 的文件可能是不一样的，需要分别执行 add 操作，但是执行一次 commit 可以提交很多文件。

10.2.3 文件跟踪管理

使用 Git 管理文件时，会经常遇到修改文件的场景。继续修改刚创建的 studygit.txt 文件，修改内容如下：

```
hello,git!
This is my second learn git!!!
```

使用 git status 命令查看仓库的当前状态，示例如下：

```
23939@DESKTOP-80LK128 MINGW64 /d/mygit (master)
$ git status
On branch master
Changes not staged for commit:
```

```
        (use "git add <file>..." to update what will be committed)
        (use "git checkout -- <file>..." to discard changes in working directory)
            modified:   studygit.txt
no changes added to commit (use "git add" and/or "git commit -a")
```

根据返回的结果得知，studygit.txt 文件已经被修改过，但是该文件还没有准备提交。有些情况下，我们记不清具体修改了文件的哪些地方，这时可以使用 git diff 命令来查看修改的内容，示例如下：

```
23939@DESKTOP-80LK128 MINGW64 /d/mygit (master)
$ git diff studygit.txt
diff --git a/studygit.txt b/studygit.txt
index 6f7d0f9..a890427 100644
--- a/studygit.txt
+++ b/studygit.txt
@@ -1,2 +1,2 @@
 hello,git!
-This is my first learn git!!!
+This is my second learn git!!!
```

diff 是 difference 的简写，表示不同。通过上面的代码可知，在文件第一行中把第二行的 first 单词修改为 second。继续将 studygit.txt 文件提交到仓库，使用 git add 命令，示例如下：

```
23939@DESKTOP-80LK128 MINGW64 /d/mygit (master)
$ git add studygit.txt
```

继续提交到仓库，使用 git commit -m 命令，示例如下：

```
23939@DESKTOP-80LK128 MINGW64 /d/mygit (master)
$ git commit -m '第二次提交 studygit.txt'
[master d3f7c74] 第二次提交 studygit.txt
 1 file changed,1 insertion(+),1 deletion(-)
```

提交后再用 git status 命令查看仓库的当前状态，示例如下：

```
23939@DESKTOP-80LK128 MINGW64 /d/mygit (master)
$ git status
On branch master
nothing to commit,working tree clean
```

通过提交结果信息，可以得出当前没有需要提交的修改，而且工作目录是干净（working tree clean）的。

10.2.4 版本回退

在实际工作中，可以对一个文件进行多次修改，但是有些情况下又想回退到之前修改的版本。本小节演示如何使用 Git 来进行版本回退。

继续对 studygit.txt 文件新增内容，内容如下：

```
hello,git!
This is my second learn git!!!
git is a great versioning tools!
```

然后使用 git add、git commit -m 命令提交到仓库，示例如下：

```
23939@DESKTOP-80LK128 MINGW64 /d/mygit (master)
$ git add studygit.txt
23939@DESKTOP-80LK128 MINGW64 /d/mygit (master)
$ git commit -m '第三次提交 studygit.txt'
[master 4e2c025] 第三次提交 studygit.txt
 1 file changed,1 insertion(+)
```

在实际工作中，可能会对文件不断地修改和提交，Git 提供了 git log 命令来查看提交的历史记录，示例如下：

```
23939@DESKTOP-80LK128 MINGW64 /d/mygit (master)
$ git log
commit 4e2c025e0c76056c601c3d51885308d7b97fefc2 (HEAD -> master)
Author: luruifeng <2393989903@qq.com>
Date:   Sun Feb 17 13:32:09 2019 +0800

    第三次提交 studygit.txt
commit d3f7c74338f4b1a0c90f27eca7918dbf863f0591
Author: luruifeng <2393989903@qq.com>
Date:   Sun Feb 17 13:16:39 2019 +0800

    第二次提交 studygit.txt
commit 0fa8d459aacf5080673bf1fff21dea0585ee267c
Author: luruifeng <2393989903@qq.com>
Date:   Sun Feb 17 13:06:43 2019 +0800

    第一次提交 studygit.txt
```

git log 命令显示从最近到最远的提交日志。由结果可以看到有 3 次提交，最近的一次是"第三次提交 studygit.txt"，上一次是"第二次提交 studygit.txt"，最早的一次是"第一次提交 studygit.txt"。如果输出信息太多，可以加上 --pretty=oneline 参数，示例如下：

```
23939@DESKTOP-80LK128 MINGW64 /d/mygit (master)
$ git log --pretty=oneline
```

```
4e2c025e0c76056c601c3d51885308d7b97fefc2 (HEAD -> master) 第三次提交 studygit.txt
d3f7c74338f4b1a0c90f27eca7918dbf863f0591 第二次提交 studygit.txt
0fa8d459aacf5080673bf1fff21dea0585ee267c 第一次提交 studygit.txt
```

一大串类似 4e2c0... 的字符是 commit id（版本号），用十六进制来表示。每提交一个新版本，实际上 Git 就会把它们自动串成一条时间线。如果使用可视化工具查看 Git 历史，就可以更清楚地看到提交历史的时间线。

Git 必须知道当前版本是哪个版本。在 Git 中，使用 HEAD 关键字来表示当前版本，即最新的提交。上一个版本就是 HEAD^，上上一个版本就是 HEAD^^，如果往上 50 个版本可以写成 HEAD~50。使用 git reset 命令回退到上一个版本，示例如下：

```
23939@DESKTOP-80LK128 MINGW64 /d/mygit (master)
$ git reset --hard HEAD^
HEAD is now at d3f7c74 第二次提交 studygit.txt
```

通过 git reset --hard HEAD^ 命令，把 studygit.txt 文件内容回退到了上一个版本，即"第二次提交 studygit.txt"。使用 cat 命令查看版本回退结果，示例如下：

```
23939@DESKTOP-80LK128 MINGW64 /d/mygit (master)
$ cat studygit.txt
hello,git!
This is my second learn git!!!
```

版本回退的结果成功。如果想回退到之前的某个版本，可以借助版本号（commit id）。每个版本都有版本号，类似 4e2c02.... 这种。回退到第一个版本，示例如下：

```
23939@DESKTOP-80LK128 MINGW64 /d/mygit (master)
$ git reset --hard 0fa8d
HEAD is now at 0fa8d45 第一次提交 studygit.txt
```

git reset --hard 0fa8d 表示回退到第一个版本，0fa8d 表示第一个版本的版本号，版本号没必要写全，只需要写前几位就可以（当然不能只写一两位），Git 会自动查找。

再次查看 studygit.txt 的版本回退结果，示例如下：

```
23939@DESKTOP-80LK128 MINGW64 /d/mygit (master)
$ cat studygit.txt
hello,git!
This is my first learn git!!!
```

10.3 Git 项目管理

Git 是分布式版本控制系统，同一个 Git 仓库可以分布到不同的机器上。当然，也可以将 Git 仓库中的所有文件上传到 GitHub 进行管理，GitHub 网站就是提供 Git 仓库托管服务的。任何人都可以建立自己的 GitHub 账号，并在 GitHub 上开源代码。

10.3.1 配置 GitHub

在使用 GitHub 之前，只需注册一个 GitHub 账号，就可以免费获得 Git 远程仓库。由于本地 Git 仓库和 GitHub 仓库之间的传输是通过 SSH 加密的，因此需要进行简单的设置。

（1）创建 SSH Key。在用户主目录下检查是否存在 .ssh 目录，如果有，再看这个目录下有没有 id_rsa 和 id_rsa.pub 这两个文件，如图 10.8 所示。

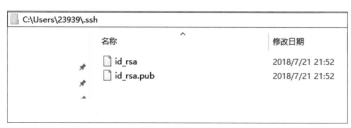

图 10.8 SSH Key 界面

如果有，直接跳过。id_rsa 是私钥，不能泄露出去；id_rsa.pub 是公钥，可以放心地告诉任何人。如果没有，打开 Git Bash 命令行，输入如下命令：

```
23939@DESKTOP-80LK128 MINGW64 ~/Desktop
$ ssh-keygen  -t rsa -C "youremail@example.com"
```

需要把邮件地址换成自己的邮件地址，然后选择默认选项并按回车键即可，示例如下：

```
23939@DESKTOP-80LK128 MINGW64 ~/Desktop
$ ssh-keygen -t rsa -C "2393989903@qq.com"
Generating public/private rsa key pair.
Enter file in which to save the key (/c/Users/23939/.ssh/id_rsa):
Enter passphrase (empty for no passphrase):
Enter same passphrase again:
Your identification has been saved in /c/Users/23939/.ssh/id_rsa.
Your public key has been saved in /c/Users/23939/.ssh/id_rsa.pub.
The key fingerprint is:
```

第 10 章 Git 版本控制工具实战

```
SHA256:tj/vloKJOCH40rnbSwuH/Uqs+W73ri7BnSybWZO1DI8 2393989903@qq.com
The key's randomart image is:
+---[RSA 2048]----+
|       . .       |
| . . o oS.       |
|. .+= E.+.       |
| ooo*X o.o  .    |
|. +0Oo o.o o     |
| .=*BB++o .*o    |
+----[SHA256]-----+
```

（2）登录 GitHub，打开"settings"中的 SSH and GPG keys 页面，然后单击"Add SSH Key"按钮，输入任意 title，在"Key"文本框中粘贴 id_rsa.pub 文件的内容，如图 10.9 所示。

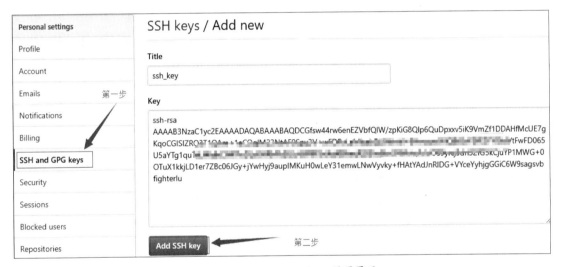

图 10.9　SSH and GPG keys 设置界面

（3）单击图 10.9 中的"Add SSH key"按钮，即可看到已经添加的 SSH key，如图 10.10 所示。

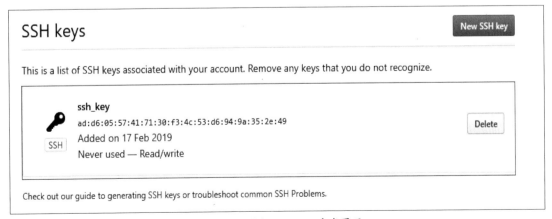

图 10.10　添加 SSH key 完成界面

10.3.2　添加远程库

使用远程仓库的目的是实现备份和代码共享的集中化管理，并且可以将远程版本库中的最新代码同步到本地，以及将修改后的代码同步到远程版本库。

前面已经在本地创建了一个 Git 仓库，也可以在 GitHub 创建一个 Git 仓库，并且希望这两个仓库进行远程同步，这样 GitHub 的仓库既可以作为备份，又可以方便其他人通过该仓库来协作。首先登录 GitHub，然后在右上角找到 "create a new repository" 选项，在弹出的菜单中选择 "New repository" 选项，创建一个新的仓库，如图 10.11 所示。

图 10.11　创建新的仓库入口界面

在弹出的界面中填入仓库名称（Repository name）、仓库描述（Description）等信息，单击 "Create repository" 按钮，如图 10.12 所示。

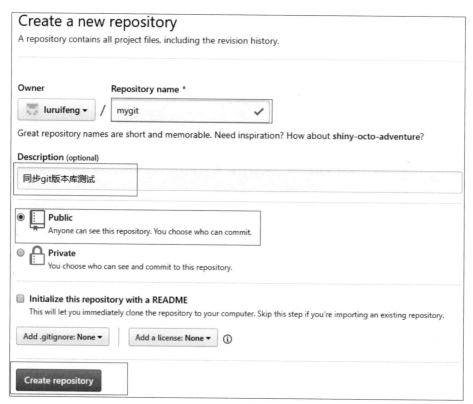

图 10.12　Github 仓库设置界面

需要说明的是,Repository name 必须要和本地 Git 仓库的名称（mygit）保持一致。创建成功后,查看结果,如图 10.13 所示。

在 GitHub 上的 mygit 仓库现在还是空的,可以从 GitHub 仓库中克隆出新的仓库,也可以把一个已有的本地仓库与之关联,然后将本地仓库的内容推送到 GitHub 仓库中。在本地的 mygit 仓库下运行如下命令:

```
23939@DESKTOP-80LK128 MINGW64 /d/mygit (master)
$ git remote add origin https://github.com/luruifeng/mygit.git
23939@DESKTOP-80LK128 MINGW64 /d/mygit (master)
$ git push -u origin master
```

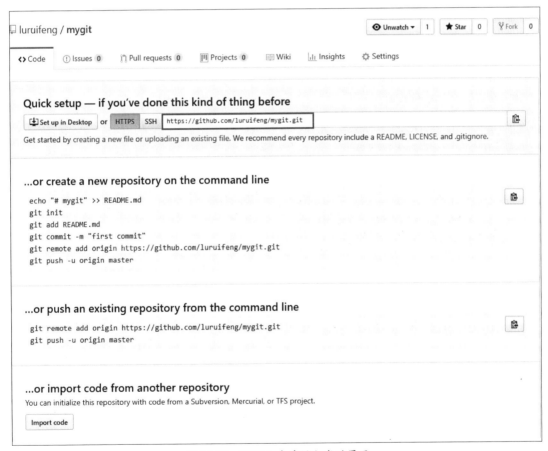

图 10.13　GitHub 仓库添加成功界面

运行命令后,弹出如图 10.14 所示界面,输入 GiHub 的账号和密码。

图 10.14　GitHub 登录验证界面

登录成功后，查看 GitBash 界面，示例如下：

```
23939@DESKTOP-80LK128 MINGW64 /d/mygit (master)
$ git push -u origin master
Enumerating objects: 3,done.
Counting objects: 100% (3/3),done.
Writing objects: 100% (3/3),277 bytes | 30.00 KiB/s,done.
Total 3 (delta 0),reused 0 (delta 0)
To https://github.com/luruifeng/mygit.git
 * [new branch]      master -> master
Branch 'master' set up to track remote branch 'master' from 'origin'.
```

将本地 Git 仓库的内容推送到远程库——GitHub 仓库中。使用 git push 命令，实际上是把当前分支 master 推送到远程。

由于远程库是空的，第一次推送 master 分支时加上了 -u 参数。推送成功后，可以立刻在 GitHub 页面中看到远程库的内容已经和本地一模一样了，如图 10.15 所示。

图 10.15　推送到 GitHub 远程库界面

之后，只要在本地进行了提交，就可以通过命令 git push origin master 把本地 master 分支的最新修改推送到 GitHub 上。

10.3.3 克隆远程库

如果实际工作中远程库中有新的内容更新了，想把远程库克隆到本地 Git 仓库中，应该怎么办呢？本小节演示从远程库克隆项目到本地。

（1）登录 GitHub，创建一个新的仓库，命名为"mygit2"，如图 10.16 所示。

（2）使用命令 git clone 克隆一个本地库。打开本地 GitBash 界面，输入 git clone 命令，示例如下：

```
23939@DESKTOP-80LK128 MINGW64 /d/mygit (master)
$ git clone https://github.com/luruifeng/mygit2.git
Cloning into 'mygit2'...
warning: You appear to have cloned an empty repository.
```

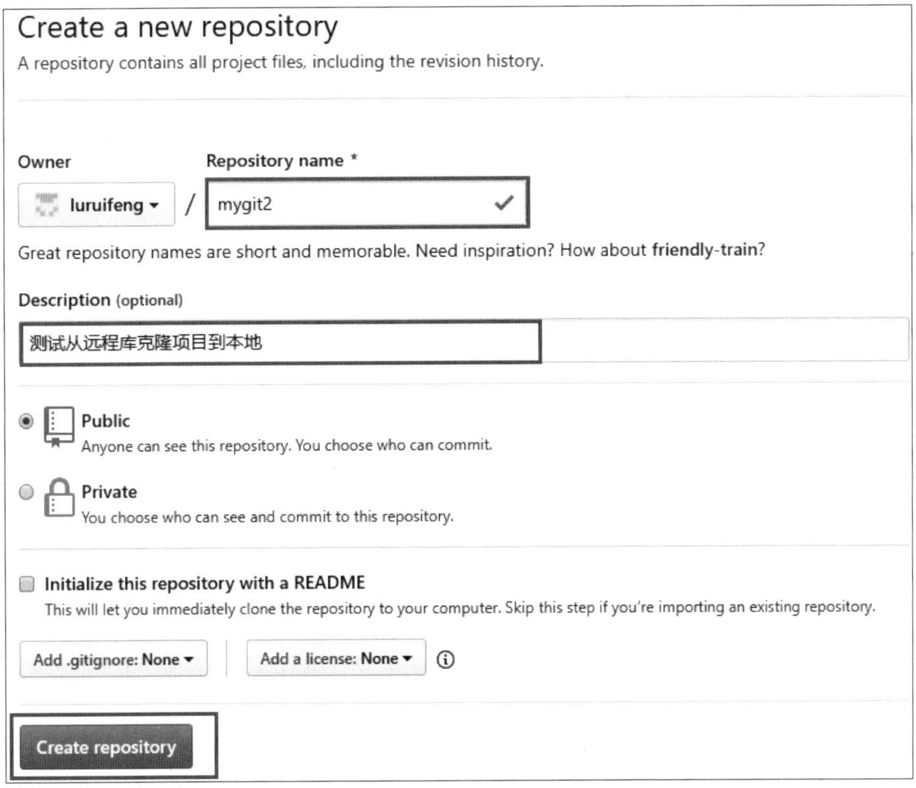

图 10.16　mygit2 仓库设置界面

返回结果中 waring 警告可以忽略，提示克隆了一个空的本地库。进入本地 mygit 仓库，查看克隆后的结果，如图 10.17 所示。

图 10.17　mygit 仓库界面

本章分享了 Git 版本工具在实际工作中的基础应用。当然，Git 的知识学习远不止笔者介绍的这些。关于 Git 的其他用法，读者可以参考其他资料。